Editor
Prof. Dr. Tosiyasu L. Kunii
The University of Tokyo
Hongo, Bunkyo-ku, Tokyo, Japan

ISBN 3-540-70002-1 Springer-Verlag Berlin Heidelberg New York Tokyo
ISBN 0-387-70002-1 Springer-Verlag New York Heidelberg Berlin Tokyo
ISBN 4-431-70002-1 Springer-Verlag Tokyo Berlin Heidelberg New York

Printing and binding: Sin-Nihon Printing, Tokyo

Many computer applications have come to have a complexity far beyond
the capability of computer hardware currently available through commer-
cial vendors. What is called the "software crisis" comes from an almost
desparate effort to fill this growing gap by software. Software engineering is aiming at easing the crisis by providing various disciplined
software tools and methodologies, and has been partially successful in
filling the gap.
This volume is the result of our effort to go beyond the software
engineering approach. Our goal is to remove the gap rather than fill it
by cost effectively building application processors in hardware.
Examples of the concepts, disciplines, methodologies, tools, architecture, design and implementation needed for such application-oriented
computers are presented by leading experts invited from related technical areas. VLSI (very large scale integration) is the hardware technology which can easily cope with any degree of complexity normally
expected in an application.
This volume focuses on VLSI-related engineering as it is used to
produce application-oriented computers. This focus is expected to give
further technical direction, integrity and uniformity to the papers.
This VLSI engineering approach can lead computer technology beyond
software engineering, and hence can contribute to overcoming the software crisis.
In Chapter 1, the authors compare software and hardware (especially
VLSI) from the viewpoint of design processes. It clarifies similarities and differences of their structures and tools, giving improved
insights into approaches beyond software engineering through VLSI. The
basic framework is given to develop advanced systems step-wise starting
from software requirement specification down to customized VLSI silicon
chip design.
Chapter 2 reports the development of various advanced architectures
which includes 5th generation computers, object-oriented architecture
and tightly coupled network architecture. The continuing rapid progress
of VLSI technology is beginning to make possible to build very large
object-oriented, highly parallel computers and special purpose machines.
Chapter 3 first addresses the problem of optimal implementation of
a class of computational processes. Specifically, it identifies computational structures that are well suited for VLSI implementation in the
form of what are called systolic algorithms. Application includes computations in pattern matching and error-correcting. Next, a class of
high speed algorithms by combinatorial circuits and their hardware
implementation methdology is proposed in terms of the VLSI algorithm
called a Bus Connected Cellular Array. The complexity theory of hardware algorithms is surveyed in connection with the present topic.
One undisputed consensus on the recent trend on computing is the
need for highly parallel computing. Central problems associated with
highly parallel computing include optimal architectures for a given
class of problems, optimal network configurations among processors and
effective methods for algorithm design and programming. Due to the
increasing complexity of VLSI circuits, design and testing problems are
becoming more and more important. There are two possible approaches to

Lecture Notes in Computer Science

Lecture Notes in Computer Science

Edited by G. Goos and J. Hartmanis

163

VLSI Engineering

Beyond Software Engineering

Edited by Tosiyasu L. Kunii

Springer-Verlag
Berlin Heidelberg New York Tokyo 1984

these problems which are discussed in Chapter 4: (1) To extend the traditional logic design and testing methods to meet the requirements of
VLSI circuits, and (2) to utilize techniques developed in the area of
software design and verification. The second approach is a good example
of a bridge between software engineering and hardware engineering.
First, the second approach is discussed in connection with two important topics: one is with parallel computation and the other is with
the logic programming language PROLOG. Graph-based design specification
of parallel computation is studied and a new method is proposed. On the
second topic, it is explained how PROLOG is used to write input and
output assertions as well as hardware specifications. Next, a practical
approach to VLSI verification and correction is briefly summarized. At
the end of Chapter 4, a new VLSI design method is proposed. A flowchart
specification is translated into a network of interconnected boxes and
local transformations are applied. This method can be used to transform
an implementation from one form into another.

Chapter 5 deals with VLSI implementation of database systems and
document image processors: (1) A hardware file system for efficient
database processing with reduction of communication cost among VLSI
circuits, (2) Highly parallel database machines focusing on search operations, (3) A picture database machine with interconnection architecture among a large number of VLSI chips and chip architecture, and (4)
Methods and requirements to produce high resolution document image
processors which exploit VLSI technology. The first part discusses an
approach in which a data storage device processes data, and a single
circuit performs various operations. Efficient processing is attained
by the principle of partitioning (division of the data to be processed
into subsets and parallel processing of these). The second part describes principles of parallel database operations, comparisons of
typical search methods, and the analysis of a search algorithm (parallel search trees). The third part presents a graph based interconnection diagram and PASCAL-like language to be used to support an extended
relational calculus for describing data operations, and an algorithm
to decompose the extended relational calculus into interconnection
diagrams. The last part develops a new dimension to utilize customized
VLSI technology. The case of a digital document image processor is in-
tensively studied, and actual examples are illustrated.

Most of the authors have been invited from among the contributors to
the 16th IBM Japan Computer Science Symposium on VLSI Engineering —
beyond software engineering — held at Hakone, Shizuoka Prefecture,
October, 1982. The editor is grateful to IBM Japan for its support in
making the updated version of important papers available for this book.

Tokyo, Japan   Tosiyasu L. Kunii, Editor

List of Contents

List of Contributors

The page numbers given below refer to
the page on which contribution bigins.

Chapter 1
Beyond Software Engineering

FUTURE DIRECTIONS FOR

VLSI AND SOFTWARE ENGINEERING

Connie U. Smith and John A. Dallen
Design Methodology Laboratory
Duke University
Durham, N.C. 27706

ABSTRACT

A structure is developed and used for the comparison of the VLSI and software design processes. The comparison reveals both similarities and differences in the two disciplines and in their evolution. Several possible implications are presented which contribute to an understanding of the processes and provide insights into appropriate directions for future research. Finally, some areas with high potential gains through technology transfer are discussed.

Introduction

Until recently the fields of software design and VLSI design have been distinctly segregated; the end products of the design process and the orientation and training of the designers have been very different. There are now, however, some compelling reasons for comparing and contrasting the two processes. Recent VLSI technological advances have dramatically increased the potential scope and complexity of VLSI designs. A VLSI engineering discipline similar to that of software engineering is essential to manage both the design complexity and the larger development projects (large teams of designers required in order to realize the new potential). Likewise, software engineers can benefit from the fresh perspective of VLSI designers. Their engineering background lends itself to the recognition of new areas in which potential gains can be realized from the technology transfer of standard engineering practices to the software domain.

There are good reasons for drawing analogies between the VLSI and software design processes. All design methodologies, irrespective of the discipline, embody the same developmental stages. A conceptual design is transformed into a physical reality by gradually refining the implementation details. The design is tested and evaluated to verify that it meets the design objectives or requirements. There are additional similarities in the two. In particular, software and VLSI design products are both computer-related objects. Frequently used software functions are now being implemented in silicon; the tools and languages used in design are strikingly similar; and many software developers are now attracted to the new VLSI domain. And finally, the evaluation of design quality during the early development stages has become an increasingly important consideration in both domains. The quality of the resulting products is often inter-related, notably so for performance characteristics.

Therefore, by comparing software and VLSI design processes an assessment can be made of the current state of the art with respect to their evolution and the relative maturity of the methodologies used in the design stages. In order to make the comparison, both design processes are mapped onto a similar structure to graphically depict both past and current design methods. The comparison reveals the similar trends in the development of design tools and methodologies. It also indicates some areas with high potential gains from technology transfer and provides some useful insights into areas for future research in both domains.

The next section develops the framework for the comparison. Section 3 then presents the evolution of the methods and the assessment of the current state of the art. Section 4 contains the implications for technology transfer and for future research. Two particularly important evolutionary developments are discussed and their future potential is considered in section 5. Finally, section 6 offers some conclusions.

2. Framework for Comparison

First, consider the stages in the design process. Chart 1 illustrates the familiar software life cycle [Bo76]. It depicts the (iterative) design process in which the system requirements at a high level of abstraction are refined into increasing levels of detail until the final, fully detailed level is implemented and enters the operational stage. The levels of abstraction should be regarded as a convenient division of a continuum rather than as discrete levels -- especially at the higher levels.

The highest level is the requirements definition. It is ideally a precise statement of the (users') requirements that the software system must satisfy, without regard to how it will be accomplished. Next is the system specification level. This is a definition of the software system proposed to satisfy the requirements. The next level is the decomposition of the software system. The system is partitioned into smaller units each of which may be designed and implemented separately. The transition from the system specification to the decomposition level usually changes the orientation from systems design to program or procedure design.

The highest three levels can be construed as the architecture layer. They embody the basic design decisions of the software system and address what will be done, generally without regard to exactly how it will be done. The next two levels, the algorithm and macro language levels, can be grouped to form the implementation layer. In it, the decisions are made on the data structures and algorithms to be used within the programs/procedures and how they will be translated into programming languages.

The lowest levels of abstraction comprise the realization layer. The assembly language level includes primitive instructions and the relative positions of data items and code. The hardware instruction level is the directly executable code. It is the end-product of the software development endeavor.

It is not necessary that a design be explicitly expressed at every level of abstraction. However, even the simplest designs must pass through descriptions within each general layer: architecture, implementation, and realization.

The design process cannot be completely characterized by levels of abstraction. At each level, a description of a design can be considered a model that facilitates the expression and evaluation of properties of the design. These properties can be differentiated into behavioral, structural, and physical categories.

Within the category of behavioral properties, it is possible to further distinguish between functional, qualitative, and execution properties. Functional behavior includes those aspects of a design which deal with its logical correctness at a given level of abstraction -- in other words, WHAT a design does. Performance is considered inherent to the functional behavior to the extent that performance affects correctness. For example, in a weapons system, a response to a critical event is required within a specified time interval in order for the software to be regarded as correct.

Additional behavioral properties address the quality of the design. Here a designer is concerned with HOW WELL a design does, while doing what it does. Factors which must be considered are: testability, reliability, modifiability, efficiency, understandability, flexibility, complexity, and implementability.

The third type of behavioral properties, execution behavior, address the dynamic aspects of the design. They concern the actions performed and address the issue of WHEN and in what order they occur. The models are generally graph-based or state-transition descriptions.

In contrast the structural category of model properties reflect the static organization of the design. It includes the hierarchical definition of components in terms of subcomponents and their connectivity. Finally, the physical model properties deal with the culmination of the design. They include the executable form of the code, the internal representation of the data structures, and the mapping to their positions in the address space of the software. These physical properties are important in that they are the necessary culmination of the design process; however, they are now typically transparent to the designer due to the prevalence of tools for the automatic transformation from structural to physical form.

A similar structure can be derived for the VLSI design process. The "VLSI life cycle" is not well established as in the case of software, but it would be very similar. The levels of abstraction are also similar, but different terminology is necessary for descriptive purposes.

The highest level of abstraction, conceptual, is generally an informal level of design description. Typically, verbal descriptions and specifications are used, augmented by charts and graphs. This level includes high level algorithmic descriptions. The architectural level is a formalization of the conceptual level description and includes block diagrams as a means of description. Typical of architectural level descriptions is PMS notation [SI74a]. The next level has traditionally been called the register transfer level of design and is usually identified with the transfer of information between entities described in the architectural level. The transition from architectural to register transfer levels is normally accompanied by a change in orientation from systems properties to concern for a particular VLSI chip design. Computer hardware design languages, such as ISP [SI74b], CDL [CH74], DDL [DI74], and their many derivatives exemplify this level.

These highest levels of abstraction can be grouped as a systems level or layer of abstraction and equate, in computer hardware design terms, to Blaauw and Brooks' architectural level [BL81] and to Siewiorek, Bell and Newell's PMS and Program levels [SI82].

The next lower level is the logic level -- the mainstay of TTL technology. Here combinational logic and state-preserving elements are detailed. It may be true, however, that this level is becoming less important in VLSI design in favor of the next level, circuit descriptions. At the circuit level, descriptions are in terms of gates and devices. Both of these levels can be categorized as being in the switching network layer. The concern here is to map system level components into physical hardware elements. Siewiorek, Bell and Newell's logic level and Blaauw and Brooks' implementation level can be considered as analogous.

The last two levels relevant to the VLSI designer comprise the geometry layer of design. At the flexible geometry level the design, as a minimum, is described in terms of a planar topology. Normally, orientations of primitives are set as well as their relative positions. However, physical sizes and absolute locations are not yet fixed. The lowest level in the design domain, also known as the layout or mask level, is that level normally required for input to the fabrication process. All characteristics are set, except possibly scale. The geometry layer equates to Blaauw and Brooks' realization level of design.

The VLSI design properties associated with each level of abstraction are also behavioral, structural, and physical. Within the category of behavioral properties a further distinction is made between functional, performance, and state transition properties. As before, functional properties address the logical correctness of a design at a given level of abstraction, that is, WHAT the design does. Performance behavior is a subset of the "qualitative" category used in the software domain. Performance has been the predominant consideration in VLSI design quality. That is, the question of HOW WELL a design does is currently equivalent to HOW FAST in the VLSI domain. The third type of behavioral property is the state-transition or procedural property. Until recently, procedural models were more common in the software domain than in hardware, but graph-based behavioral models are being proposed to consider the dynamic tracing of what and how a design does.

As with software design, structural model properties include the hierarchical definition of components in terms of subcomponents and connectivity information. Physical model properties address the actual layout of the design on a chip area.

The next step in the comparison of the design processes is the taxonomy of existing models. Charts 2 and 3 depict the software and VLSI domains, respectively, with the levels of abstraction vertically and the model properties horizontally. A cross-section of models are considered and included are primarily model description languages as opposed to modeling systems or tools. Each is placed on the chart according to the primary level of abstraction and model property that it addresses. The placement is derived by studying the primitives of the language and ascertaining the essence of the information contained within the primitive. If it is commonly applicable to multiple levels or properties, its scope is indicated by arrows or boundary-lapping descriptions. While some languages can be considered to possess properties from several of the categories and can often be successfully used at several levels of abstraction, most are heavily oriented to one primary model property and to one or two levels of abstraction.

3. Comparison

3.1. Design processes

Having completed the framework, we can now begin to make comparison of the design processes. Charts 2 and 3 are quite similar in appearance. Columns and rows have analogous meanings even though the names may differ. In both disciplines, the design process can be seen as a progression from the top-left to the bottom-right of the chart, taking one or more routes through the various levels. At each level, a description language is used to represent the design. The difficulty lies in making the transitions from one level of abstraction to the next, and this mapping process, whether manual or automatic, constitutes the art and science of design. In both cases, the process is iterative in nature, requiring a return to design at higher levels of abstraction when problems arise at lower levels.

Not only are the design processes similar, but also the important elements of the design methodologies are common to both domains. These elements, shown in chart 4, include design principles, design objectives, and techniques for the testing and evaluation of a design. Design principles are proven techniques used in the formulation of a design which help to achieve the objective of the design. The objective of the design process is a completed description of the design (physical information at the lowest level of abstraction) which must be correct and "good". Assurance of correctness requires functional behavior information and assurance of quality requires metrics to characterize quality based on combinations of structural, behavioral, and physical properties.

Design test and evaluation procedures provide the information necessary to determine whether or not the objectives have been met. Test and evaluation techniques in the VLSI domain have been dominated by simulation methods. Analytic analysis techniques could theoretically provide more timely information on the status of a design than is usually possible with simulation. In the software domain, analytic techniques are being developed and used for evaluation. Testing is still largely conducted through repetitive execution of the software under varying conditions. Formal verification techniques are desirable in both domains, but are as yet impractical for very complex designs.

There is a discrepancy between the behavioral properties of performance in the VLSI domain and quality in the software domain. This is perhaps due to the relative youth of VLSI and to the fact that designs are currently dominated by technological constraints. If this is true, one can expect that the notion of quality in the VLSI domain may soon become closer to that of software. Quality in the software domain includes performance as well as reliability, modifiability, and other attributes as shown on chart 4.

There are two notable differences in the two design processes depicted on charts 2 and 3. First, there is a considerable amount of model transparency in the software domain. That is, the transformation of designs from structural/algorithmic or macro language specifications into their machine instruction/physical form is typically transparent, due to the prevalence of automated tools such as compilers, linkers, and loaders. Second, the labor intensive efforts occur at different stages of the development process. In the VLSI domain, manual transformations are required in the lower-right portion

of the chart. This is extremely labor intensive due to the complexity
of the designs and the critical chip area considerations. In
software, the labor intensive efforts occur between the
conceptual/functional and algorithms/structural transformations.

Additional comparative information can be obtained through consideration of the evolution of the design process.

3.2. Evolution of design processes

The evolution of the design processes has many similarities.
Initial breakthroughs were made in a "bottom-up" fashion. Tools and
techniques first addressed the transformation of design representations in the lower-right corner of the charts. VLSI design tools have
been physically oriented and based on descriptions at the geometry
level. This state of the art is reflected in the prevalence of design
tools which use CIF[Me80], X-Y Mask, and other low level descriptions
as the primary building block. Then, design tools which attempted to
use descriptions at the flexible geometry level were actively pursued,
such as William´s Sticks[Wi77] and Johannson´s early form of Bristle
Blocks[Jo79]. More recently, attempts are being made to use circuit
level descriptions as the building blocks, exemplified by Rosenberg´s
ABCD[Ro82] and Cardelli and Plotkin´s Sticks and Stones[Ca81]. In all
cases of this bottom-up evolution, the orientation has been physical
and structural.

The evolution of the software development process is similar to
that of VLSI. In the 1950´s attention was focused on physical properties of software at the assembly language and hardware instruction
levels. Coding and memory layout were major concerns and quite time
intensive activities. Subsequently, higher level languages such as
FORTRAN and COBOL relieved many of the coding duties, but memory was
still a primary concern. More recently, algorithmic languages such as
ALGOL and Pascal have become the building blocks.

Early in the evolution of both disciplines, focus was (is) on
time and space requirements. This was primarily due to physical constraints imposed by the state of the technology; both memory and chip
area were limited. In the software domain, technological advances
lessened the physical memory constraints and thus made automated (but
less efficient) tools more tractable. Compilers, and later operating
systems, had a dramatic effect on productivity and thus the scope of
the designs that could feasibly be implemented. It appears that a
similar state will be achieved in the VLSI domain.

It is interesting to note that there is considerable resistance
among expert VLSI designers to the development and use of fully
automated layout systems since they will be inherently less efficient
than the custom layouts. Similar resistance was encountered when FOR-
TRAN was first introduced. Extra attention was devoted to optimizing
the code produced by FORTRAN compilers in order to mitigate this
resistance. Such super-optimization has not been included in later
versions of compilers, once initial acceptance of the tool was
achieved.

Following the resolution of the hardware technology bottleneck in
the software domain was a dramatic increase in the ambition and complexity of the systems designed. Software complexity has increased to
the point where the system definition phase has now become crucial. A

severe problem that arises as a result of this increased complexity is the testing, verification, and analysis of the resulting software. This problem has already been encountered in the VLSI domain and its severity is likely to increase due to the impending dramatic increase in VLSI complexity. The current technique of design verification via the extraction of behavioral descriptions from lower level physical descriptions will no longer be feasible.

Paralleling the evolution of design processes originating at the lower levels have been attempts to extend system level techniques to VLSI design. The basis of these attempts has been through the adaptation of common register transfer level descriptions. Most of these descriptive languages are heavily oriented towards computer hardware design, such as ISP, CDL, AHPL, RTL and so on. Most of these languages are functionally oriented, however, requiring parallel design development in the structural and physical domains. The advantage of functional descriptions, of course, is that some analysis of a design is made early in the design process. Unfortunately, unless equivalence can be proven between the behavioral, structural, and physical models at any level of abstraction, final design verification and analysis must still rely on testing and evaluating lower level descriptions.

Bridging this gap from the system-behavior domain to the physical-geometry domain is the essence of the design problem. Ideal solutions include the search for the mythical silicon compiler, the definition of some standard notation or algebra of design as is used in more established disciplines, and the invention of some universal description language appropriate for behavioral, structural and physical description at all levels of abstraction. On the other hand, most practical design systems existing today rely on a database approach to design in which several different descriptions are used and a great deal of human interaction is required to effect the mapping between levels. It is obvious from the literature, however, that the problems of design verification and analysis are getting worse and not better as the number of devices on a chip keeps growing.

Some differences in the evolution of the design processes are also noticeable. In the software domains precise, formal specification languages have been proposed for requirements definition and system specifications. There is a trend toward evaluating the software at this level to access its quality and thus ensure that an appropriate design is selected before implementation begins.

The consideration of formal languages for design descriptions at the systems level of abstraction has occurred comparatively early in VLSI evolution. Design languages for software have been formulated specifically for software development problems. The most recent example has been the development of ADA. Many higher level VLSI languages, on the other hand, have been adapted from other domains. Examples are the hardware description languages, ISP, CDL, DDL, and languages derived from software programming language constructs, such as MODEL.

Similarly, silicon compilers for the automatic transformation of designs from system level descriptions to the geometry level are also being studied earlier in the VLSI evolution than the counterpart automatic program generators in the software domain. Early attempts at automatic program generators actually resulted in slightly higher level programming languages which mapped to the next level of abstraction and then only in restricted application domains. Advances were made one step at a time rather than in quantum leaps.

The software design process has evolved to the state that current research efforts are focused on managing complexity and thus improving productivity at the highest levels of abstraction. In the VLSI domain, the research efforts are not concentrated on the same area, but are instead spread more uniformly across the spectrum.

4. Implications

What can we learn from this comparison of the two disciplines? First, it is apparent in the comparison that the design path taken from the functional-concept/requirements definition level to the physical-geometry/machine level is much more linear in software design than in VLSI design. Currently, a VLSI designer must simultaneously use parallel languages in several domains during the process. This is rarely true in software. This difference also reflects the fact that in software engineering the less profitable paths have been abandoned and the more profitable ones emulated. This winnowing process has just started in VLSI.

Also apparent is that VLSI designers may be able to predict, from the experience of software designers, future trends in the evolution of VLSI design tools. Just as memory layout became less important, perhaps as chip space becomes less critical, less concern for physical layout will be observed in VLSI design. Just as the higher-level programming languages have evolved as the basic building blocks of software engineering, so perhaps will higher-level switching network constructs (not necessarily logic gates) become the building blocks of VLSI design. The efficiencies that are currently gained by designing at the circuit level (and below) may eventually fall into disuse as has programming at the assembly language level. However, we should take note of the fact that the majority of industry software systems are still implemented in COBOL (and even FORTRAN) even though algorithmic languages are generally more suitable. Industry has become "locked in" due to the vast amount of existing code (and expertise) that must be used. The same could happen in the VLSI domain.

Less comforting conclusions can be postulated, unfortunately. Major problems in software engineering have yet to be solved, indicating that similar problems may thwart efforts in VLSI design. In particular, the process of design evolution from the system level to the switching-network level has proved difficult to understand and control. Software compilers, it should be remembered, only map from the programming level to the machine level. This bodes poorly for VLSI designer efforts to produce silicon compilers which map automatically from system-level descriptions to the geometry level. It suggests that the most fruitful research in the near term is in areas that will lead to automated tools for transforming switching-network level descriptions into fixed geometry descriptions. If, on the other hand, effective system-level silicon compilers are found, perhaps software designers may be able to learn something from this.

Both disciplines face similar problems in the area of design verification and analysis. Reliance on simulation tools (test-runs in software) relegates the evaluation process to the late stages of the design process. The verification process is also slow and consumes a significant portion of the overall design effort.

Key to the timely evaluation of a design is in early measurement against design objectives. While issues of design quality are recognized as important, no real place has been found for insuring design quality in the design process. If real progress is to be made in VLSI and software engineering, both function AND quality will have to be coped with early in the design process. This dictates the need for a better balance in the use of proofs, analysis and simulation in support of both software and VLSI design.

Many custom chips are being designed to perform black-box functions in large computer systems. Functions which were previously implemented in software, such as floating point arithmetic units, have already been implemented in VLSI, and such functions as speech synthesis and device scheduling are being considered for implementation. This will presumably remove software bottlenecks; however, a new problem is introduced. The effectiveness of the new chip will be limited by the data-flow rates into and out of the chips. This bottleneck must be resolved if maximum effectiveness is to be achieved with the new technology.

If an analogy is to be drawn between the software and VLSI design processes, one must consider the issue of maintenance in software systems. Currently sixty percent of the cost of software systems is in the maintenance stage. What is the equivalent in the VLSI domain? Software maintenance consists of three primary activities: major enhancement, minor enhancements, and correction of latent errors. On the surface, it appears that there is no equivalent in the VLSI domain since chips are not "patched" once they are fabricated. However, as can be seen in chart 1, software maintenance requires backtracking to an earlier step in the development process. The analogy in VLSI occurs when there is a need to construct a chip with slightly different functionality or improved performance. It should be possible to return to an earlier design stage and to modify the previous design representation at that stage. Unfortunately the tools available currently limit the extent to which this can be done. The only feasible modifications at this time are minor enhancements (typically speed) and error corrections. Major enhancements to designs typically require an entire redesign. In order to significantly improve productivity, new tools are needed to facilitate modifiability.

5. Technology Transfer

To date, the technology transfer has primarily been from the software engineering domain to the VLSI domain. This is because the VLSI field is relatively new and because the migration of designers has been in that direction. It is now time to begin looking for technology transfer in the other direction. Some areas that appear promising in the near term are the consideration of VLSI design principles, the use of simulation as a supplemental test/verification tool, and the development of an integrated computer aided design system. The VLSI design principles of uniformity, regularity, and symmetry facilitate the design process. These principles are typically used in other engineering disciplines as well. Uniformity is the use of components that are uniform in size and shape. Regularity applies to components that are designed once and replicated many times on the chip. This reduces both the number of components that must be custom designed and the associated implementation and verification problems. Another aspect of regularity is design consistency: the use of similar design structures in different components. An example is the designing of components such that data lines run horizontally and control lines run vertically. Symmetry refers to the shape of the components.

It allows additional flexibility in the physical layout of components. Use of these principles not only facilitates the manual implementation process, but also will enhance the quality of the designs produced by future, automated layout tools. The benefits to the software domain are in the areas of producing reusable code, facilitating more automated tools for use in the design process, enhancing maintenance and modifiability, and improving quality and performance.

In the area of test and evaluation, no single tool will ever suffice. Simulation of design functions at a high level of abstraction offers an additional mechanism for detecting and correcting errors at early developmental stages. While it is true that simulation is currently a cumbersome method to use for validation of large, complex VLSI designs, it has been effective at detecting and correcting design errors. It could be a viable method with more automated support at lower levels of abstraction and with extensions for application to higher levels of abstraction.

Most current software approaches to the development of automated tools to support the design process concern collecting useful tools for a "toolkit" or "workbench" While each tool is individually useful for a particular task, the integration of the tools is not a primary consideration. A more fruitful approach is to consider the development of an integrated system to support the design process as it passes through each phase. Generally, computer aided design systems are built around a central data base of design information. Tools are successively invoked to facilitate the mapping of the design between levels of abstraction and the evaluation of model properties at each level.

In the long term, the silicon compiler may have a significant impact on the software engineering field. It is an ambitious project and should yield additional insights into methods necessary for automatic generation of programs or systems of programs.

There have been two additional developments in the software domain, that have had significant impact on the design processes, that one must consider for transfer to the VLSI domain. They are presented in the following sections.

5.1 Operating systems

Operating systems have been a major development in the software field. They have tremendously affected both the productivity of designers and the magnitude and complexity of systems that can be feasibly implemented. What, then, would be the equivalent silicon operating system? In order to answer this question the structure and functions of an operating system will be reviewed.

The operating system can be viewed as a set of layers each of which provides an abstract machine with increased capabilities over that of the underlying layers. The abstract machine at each layer relieves its users of a dependence on the implementation details of the underlying layers.

The innermost layer, the kernel or nucleus, encapsulates the hardware dependent aspects of the operating system. It includes interrupt handling, device drivers, and concurrency control primitives. The next layer adds memory management capabilities which relieve users of concerns about memory protection and, generally, of size constraints on code. The next layer incorporates file handling capabilities. Other features such as security, data management, accounting, etc., are added in subsequent layers.

Next, consider the primary productivity benefits of operating systems. They enable sharing of the computing facilities among large numbers of users in a manner (usually) transparent to each user. They also relieve users of redundant coding of commonly used functions. The global throughput is improved, although individual (local) performance is occasionally sacrificed.

On the surface, there is no direct analogy in VLSI. There are, however, aspects of operating systems which may be applicable to the VLSI domain in the near future, and which may achieve the same productivity gains that were achieved in the software domain.

The goal of the mythical silicon operating system (SOS) is to improve designer productivity by allowing large numbers of designers to "share" (or work on) the same chip without concern about the effect on other designers, relieving designers of concerns about physical limitations, and eliminating the need to re-implement commonly used functions. A primitive SOS can be constructed by providing capabilities for communication between the design entities or components, handling timing (synchronization) constraints, handling data flow onto and off of the chip, and assigning design entities to physical chip area.

The kernel of the SOS provides the handling of data flow onto and off of the chip. The physical constraint on the number of pins and the multiplexing of pin signals (if necessary) are encapsulated into the kernel. Pin assignments could either be made in the kernel or in the equivalent "file (data) handling" layer.

The kernel also contains communications and synchronization primitives. Communication between design entities can be handled either by direct connections (wires) between components or by a "mailbox" capability which temporarily stores output from one component until needed (requested) by another. Synchronization can be provided by the SOS when multiple components must be activated simultaneously or when one component must be activated when a specified state is reached. These capabilities relieve designers of a need to know the relative speed of design components.

Designers of components work with virtual real estate. With the recent technological trends, the number of devices within a single designer's real estate could easily equal or exceed the number of devices previously contained on an entire chip. There may, however, be constraints on the maximum amount of virtual chip area allocated to a design component and guidelines for its aspect ratio. The external interfaces are also defined, but the designer need not be concerned with the component's physical location, its position relative to components it interfaces with, or their relative speed. The chip area manager is responsible for assigning (placing) components to physical areas on the chip and the communication (routing) between components. Exceptional situations can be handled by allowing the option of assigning virtual chip areas to specific locations on the chip (a virtual=real analogy).

The use of an SOS, of course, results in a chip inherently less efficient than a custom-designed chip. It does, however, have many advantages. It has potential for significantly improving the

productivity of designers and reducing the level of expertise required
for designing chips of this type. It can provide support for novice
designers and enable them to design chips that perform useful functions in a shorter period of time. It achieves these goals without
constraining the architecture of the chip as in the case of PLA's and
standard cells. Even though the efficiency of chips developed with an
SOS would not be as good as that of custom-designed chips, it would
facilitate the implementation of functions previously performed in
software, thus there would be a net gain in performance. Critical
constraints on chip area and performance unfortunately preclude the
current use of an SOS; however, technological advances are rapidly
being made and the SOS may be feasible in the near future. As in the
software domain, it could offer greater potential for productivity
gains than the automation of the design process itself as in silicon
compilers.

The silicon operating system is proposed as a development aid
targeted towards the switching network and geometry layers of abstraction and provides transparency between the structural and physical
design properties. The next topic addresses the system layers (and
those lower) and concerns the qualitative (in software) and performance (in VLSI) properties.

5.2 Quality engineering

Quality (or value) engineering is a standard engineering practice
for the evaluation of proposed designs at early developmental stages
and the selection of the most appropriate designs to be implemented.
The evaluation process then continues throughout the implementation
process to ensure that detail decisions maintain the desired quality.
It enhances the quality of the resulting product and increases
designer productivity by insuring that efforts are appropriately
invested.

There is currently no general methodology for assessing the quality of a software (or a VLSI) design. There has, however, been a
recent development in the software domain, a performance engineering
discipline, with potential for technology transfer to both the VLSI
domain and to other aspects of software quality. Performance
engineering embodies a methodology for the evaluation of the performance of software at early design stages before coding is begun. The
evaluation can be accomplished easily, in a very short period of time
(interactively). It can be used to select appropriate designs for
implementation, identify critical components with respect to performance, quantify configuration requirements, and to assess the performance impact of changes made during the development process.

The performance analysis at the architectural (or system) level
is accomplished by first gathering the necessary performance specifications: workload, environment, software structure, and resource
requirements. These are mapped onto an appropriate representation of
the dynamic execution behavior: software execution graphs. Graph
analysis algorithms are used to obtain best case, worst case, and
average case response times as well as distributions and variances.
Data dependent execution behavior is explicitly handled. The execution graphs are later mapped to queueing network models for the
evaluation of the computer configuration requirements, the effects of
concurrent execution, and analysis of other advanced design concepts.

The evaluation is iterative; design modifications and refinements can easily be incorporated and their effect analyzed. The evaluation continues throughout the design process to ensure that the resulting product meets performance goals.

It has been demonstrated that the analysis techniques can accurately derive performance metrics when accurate specifications are available. It has also been demonstrated that sufficient information is available at early design stages to accurately predict resource usage patterns and to identify performance problems before coding is begun. It is an effective technique that requires minimal effort and has the potential for tremendous performance improvements without time consuming, effort intensive tuning projects often required to correct performance problems after software is implemented [SM82].

It is logical to consider the transfer of performance engineering to the VLSI domain. It has already been successfully used to evaluate high level applications software, intermediate level data management software, low level operating system software, and even hardware architecture. The next logical level to address is VLSI. Further, performance is currently a critical part of VLSI designs: even more important than the software counterpart.

Simulation has previously been used to evaluate performance characteristics of VLSI designs. It is, however, a labor intensive effort and can only be applied at later stages in the design process. Approximate analytical techniques could be employed at earlier design stages to consider issues such as the appropriate partitioning of designs and the assignment of devices to chips and at later stages to evaluate alternative designs for devices on a chip.

The analysis techniques would be similar to those used for software. The specifications are analogous: the typical functions to be performed (workload), the execution environment of the chips, the structure of the design, and estimates for timings (resource requirements). Graph models are an appropriate representation for VLSI designs; therefore, similar algorithms can be used to derive timing information. Additional models will be required, however, to analyze highly parallel designs or those with critical synchronization requirements. Petri nets models should be investigated as an alternative to the queueing network models used for software.

The potential benefit of performance engineering to the VLSI domain is apparent. Research is required on the appropriate models for analysis of parallelism and synchronization, on the derivation of the necessary specifications at high levels of abstraction, and the incorporation of the techniques into the design process. Similar extensions are required to extend the notion of performance engineering to include other aspects of (software) design quality. Additional metrics which accurately assess the quality of a proposed design are a fundamental requirement. Models for the calculation of the design metrics can then be derived and applied to evaluate design quality prior to implementation.

6. Conclusions

On the surface, the VLSI and software design processes are quite dissimilar since one produces a tangible product while the other does not, and since one is concerned with electrical properties of materials while the other is not. Upon closer examination, however, the processes can be mapped onto a common framework for analysis and comparison. This, combined with consideration of their evolution, lends insights into the likely outcome of current research, appropriate directions for further research, and areas with high potential for increasing designer productivity and product quality through technology transfer. Examples are described for further consideration: a silicon operating system and disciplines for both silicon performance engineering and quality engineering.

References

[B181] G.A. Blaauw and F.P. Brooks, Computer Architecture, preliminary draft (1981).

[Bo76] B.W. Boehm, "Software Engineering," IEEE Transactions on Computers, Vol. C-25(12) (December 1976).

[Ca81] Luca Cardelli and Gordon Plotkin, "An Algebraic Approach to VLSI Design," pp. 173-182 in VLSI 81 Very Large Scale Integration, ed. John P. Gray, Academic Press, Inc., New York (1981).

[Ch74] Y. Chu, "Introducing CDL," Computer, Vol. 7(12), pp. 31-33, IEEE (December 1974).

[Di74] D.L. Dietmeyer, "Introducing DDL," Computer, Vol. 7(12), pp. 34-38, IEEE (December 1974).

[Jo79] D. Johannsen, "Bristle Blocks--A Silicon Compiler," in Proc. of the 16th Design Automation Conference (1979).

[Ma80] M. Marshall and L. Waller, "VLSI Pushes Super-CAD Techniques," Electronics, pp. 73-80 (July 31, 1980).

[Me80] Carver, Mead and Lynn Conway, in Introduction to VLSI Design, Addison-Wesley Publishing Company (1980).

[Ro82] J. Rosenberg and N. Weste, A Better Circuit Description, Technical Report 82-01, Microelectronics Center of North Carolina (1982).

[Si74] D. Siewiorek, "Introducing PMS," Computer, Vol. 7(12), pp. 42-44, IEEE (December 1974).

[Si74] D. Siewiorek, "Introducing ISP," Computer, Vol. 7(12), pp. 39-41, IEEE (December 1974).

[Si82] D.P. Siewiorek, C.G. Bell, and A. Newell, Computer Structures: Principles and Examples, McGraw Hill Book Company (1982).

[Smi82] Connie U. Smith and J.C. Browne, "Performance Engineering of Software Systems: A Case Study," Proc. AFIPS National Computer Conference, Houston (1982).

[Wi77] John D. Williams, STICKS--A New Approach to LSI Design, Master's Thesis, Massachusetts Institute of Technology (June 1977).

16

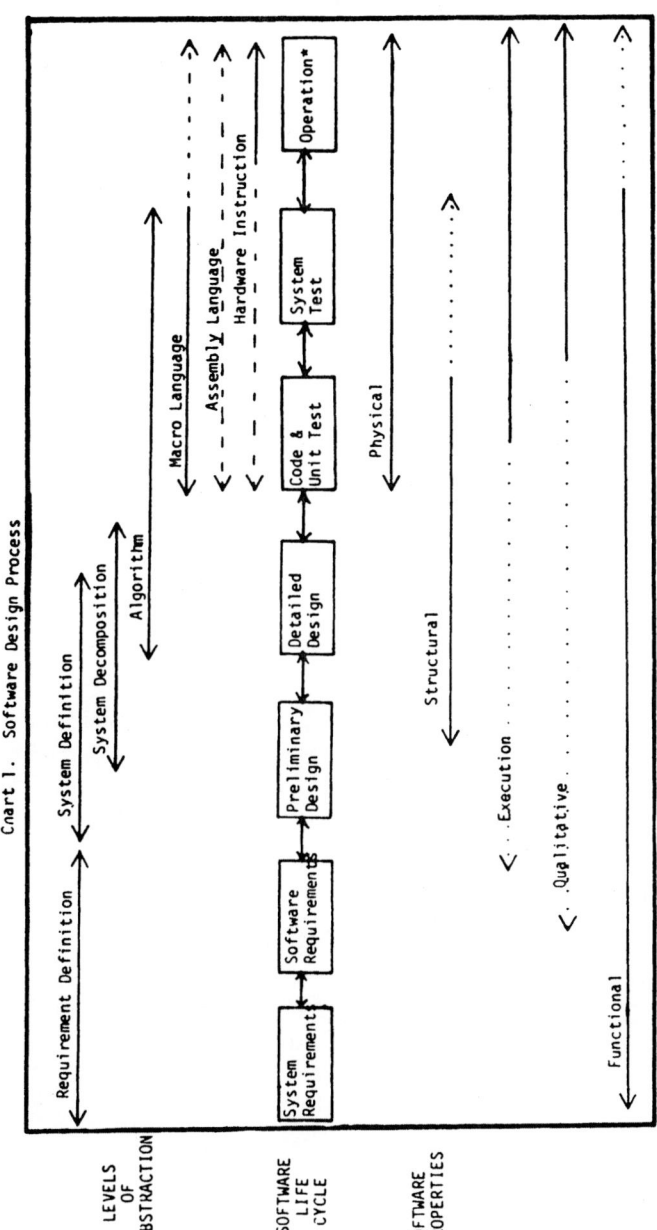

Chart 1. Software Design Process

Key: ------ applicability is usually transparent
     ...... limited consideration of these properties
     _____ strong applicability

* Note: Maintenance is not explicitly included since it can actually be defined
  by subsets of the above life cycle. That is, major enhancements = entire life
  cycle, minor enhancements = detailed design through operation, and latent error
  correction = code and debug through operation.

Chart 2

LEVELS OF
ABSTRACTION   SOFTWARE DESIGN

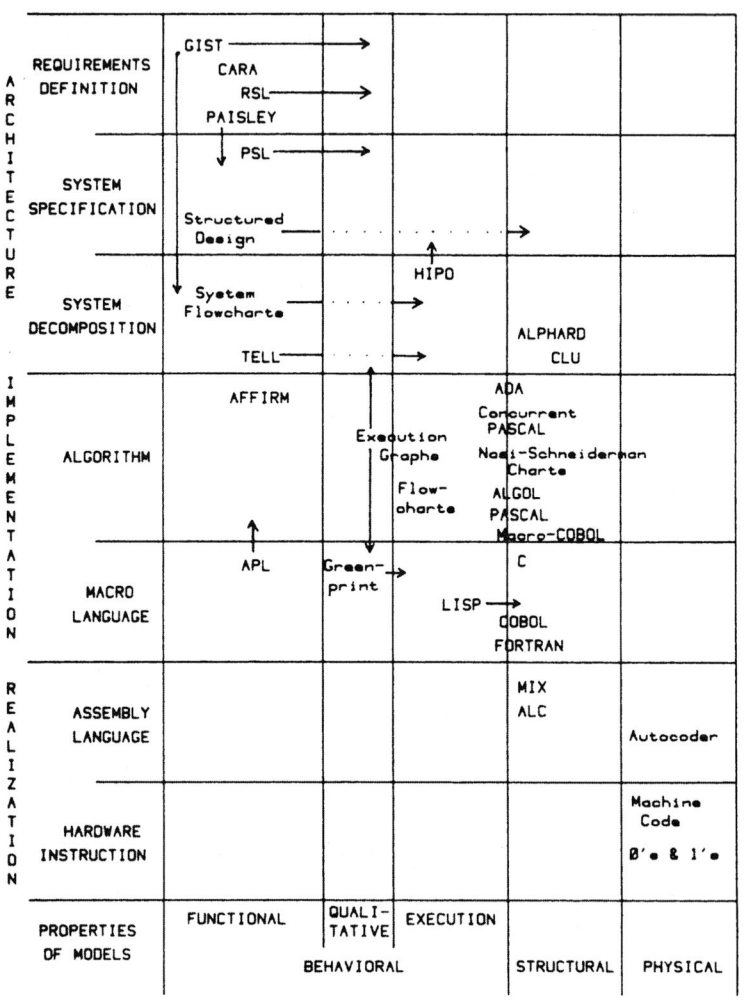

Chart 3

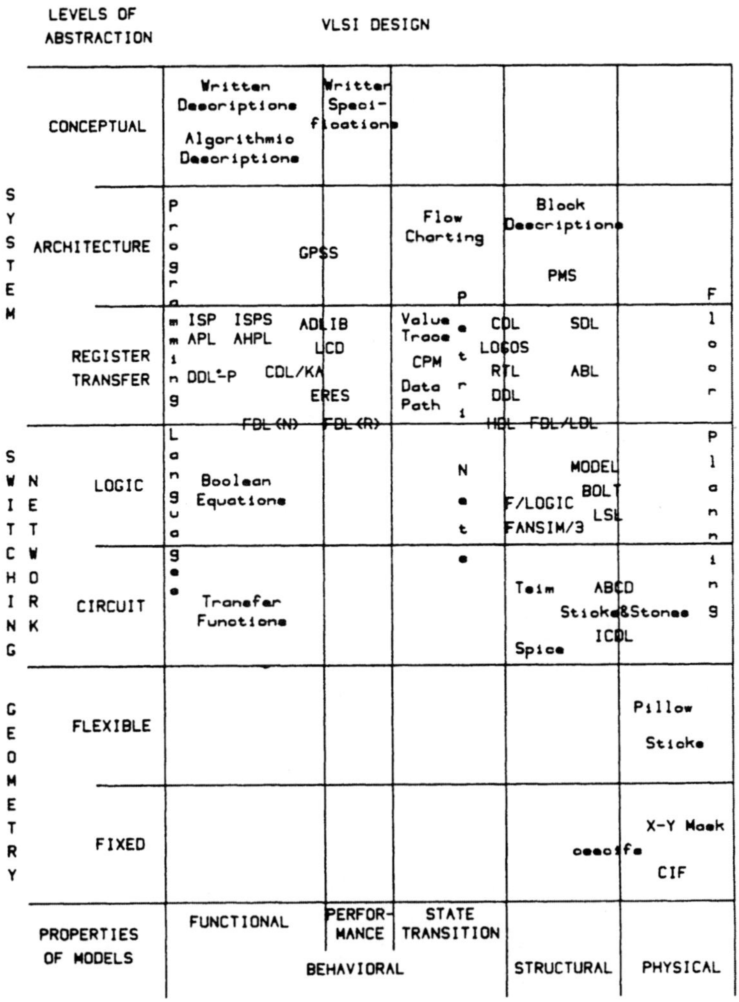

LEVELS OF ABSTRACTION — VLSI DESIGN

Levels of Abstraction	FUNCTIONAL	PERFORMANCE	STATE TRANSITION	STRUCTURAL	PHYSICAL
CONCEPTUAL	Written Descriptions / Algorithmic Descriptions	Written Specifications			
SYSTEM — ARCHITECTURE	Programming... GPSS		Flow Charting	Block Descriptions / PMS	Floor...
SYSTEM — REGISTER TRANSFER	ISP ISPS / APL AHPL / DDL-P CDL/KA / ADLIB LCD ERES / FDL(N)	FDL(R)	Value Trace / CPM / Data Path / Petri	CDL SDL / LOGOS / RTL ABL / DDL / HDL FDL/LDL	Floor Planning
SWITCHING / NETWORK — LOGIC	Language... Boolean Equations		N • t	MODEL / BOLT / F/LOGIC LSL / FANSIM/3	Planning
SWITCHING / NETWORK — CIRCUIT	Transfer Functions		•	Teim ABCD / Sticks&Stones / ICDL / Spice	ing
GEOMETRY — FLEXIBLE					Pillow / Sticks
GEOMETRY — FIXED				ceeoife	X-Y Mask / CIF
PROPERTIES OF MODELS	FUNCTIONAL (BEHAVIORAL)	PERFORMANCE (BEHAVIORAL)	STATE TRANSITION (BEHAVIORAL)	STRUCTURAL	PHYSICAL

Chart **4.** ELEMENTS OF DESIGN

PRINCIPLES OF DESIGN

- Hierarchical Decomposition
- Abstraction
- Hiding
- Localization
- Uniformity
- Regularity
- Completeness
- Confirmability
- Simplicity
- Symmetry

OBJECTIVES OF DESIGN

- Design Description
- Design Correctness
    > Feasibility
    > Equivalency
    > Consistency
    > Performance
- Design Quality
    > Testability
    > Reliability
    > Modifiability
    > Efficiency
    > Understandability
    > Flexibility
    > Complexity
    > Implementability

TEST AND EVALUATION OF DESIGN

- Proofs
- Analytic Analysis
- Simulation

Chapter 2

Systems and Machine Architecture

An Object-Oriented, Capability-Based Architecture

J. C. Browne and Todd Smith
Dept. of Computer Science and the Computation Center
The University of Texas at Austin
Austin, Texas 78712

1.0 OVERVIEW

This paper defines and describes a capability-based architecture for realizing object-oriented programming. All addressing is in terms of objects. A conventional linear address space is not visible even at the instruction level. The proposed architecture is based on making a complete separation of the concepts of physical addresses and unique identifiers. This separation allows a complete virtualization of a linear address space into an object-oriented address space. The addressing architecture associates with each capability a selector. A selector is a pointer to the physical address of the current component of each data object. The second major concept element in the architecture is to include in hardware a local address space of objects. The address space of an execution unit is defined entirely in terms of a nickname table which holds the capabilities and selectors for the objects known to the execution unit. Addresses of objects are realized only in the context of a local execution environment. A nickname table corresponds closely to a hardware realization of the "local name table" of the HYDRA [Wulf, et al, 1981] operating system. The third major concept is the use of "tags" to type the contents of memory.

The proposed architecture gives practical and implementable solutions to the small object [Lanciaux, et.al., 1978] [Gehringer, 1979] problem and the problem of addressing of composite objects. It provides a convenient solution to the seal/unseal problem [Gligor, 1979] for data objects thus providing a basis for passing extended types (user defined types) as parameters. It provides a unification of primitive and extended types. The user interface even at the assembly language level is entirely expressed in terms of objects. Addresses do not appear in code. The use of the nickname table also eliminates the use of capabilities directly in instructions.

The subsequent sections of the paper cover motivation for object-oriented architectures, a detailed description of the addressing modes of the architecture and a brief discussion of implementability and further requirements for an effective object oriented system including object-oriented secondary memory devices.

2.0 MOTIVATION

The end user of a computer program usually formulates his/her problem in terms of logical objects. This view of a problem is then translated to the linear address space realized by conventional computer systems through the transformations shown in Figure 1. The complexity of modern software systems arises to a large degree because of the multiple mappings required to go from the logical objects perceived by an end user to the linear address space realized by conventional hardware. Having all levels of software expressed in terms of objects will greatly simplify the software.

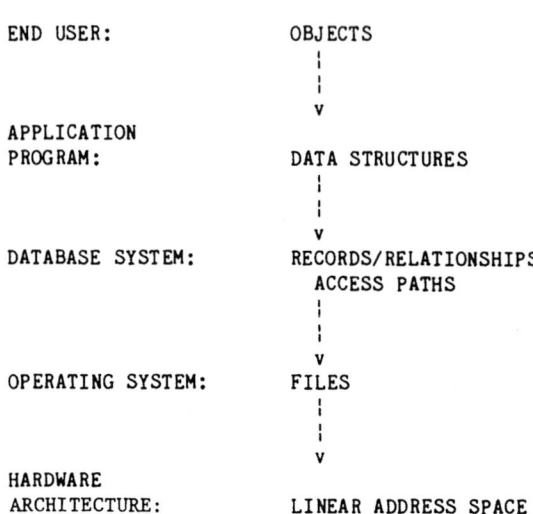

```
END USER:                OBJECTS
                            ¦
                            ¦
                            v

APPLICATION
PROGRAM:                 DATA STRUCTURES
                            ¦
                            ¦
                            v

DATABASE SYSTEM:         RECORDS/RELATIONSHIPS
                         ACCESS PATHS
                            ¦
                            ¦
                            v

OPERATING SYSTEM:        FILES
                            ¦
                            ¦
                            v

HARDWARE
ARCHITECTURE:            LINEAR ADDRESS SPACE
```

FIGURE 1: VIEWS AND MAPPINGS FROM USER TO PHYSICAL MEMORY

Experience suggests that efficiency will be hard to realize unless the hardware architecture also directly implements objects. The thesis upon which this work is based is that a complete object-oriented architecture will be a major step towards simpler and more reliable and more efficient software systems.

The first major decision is that the mechanism for the support of objects will be capability based [Fabry, 1974]. The concept of capabilities will be integrated with that of tags [Feustal,1973] to provide a compact and economical organization for the support of objects. Capabilities are a means of defining objects and mapping objects to a realizable space. Capabilities provide a natural implementation of locality since the address space of a procedure or a process can be completely specified by the set of capabilities placed in the address space. A capability based implementation of objects has the basic merit of separating structure from values. This separation is essential for both virtualization and dynamic structuring. References to objects passing through capabilities can be controlled and structured. Capabilities provide a natural mechanism thus for the implementation of abstract data types which are prototypical objects.

The next major principle is that of uniformity and transparency of memory. Only objects will be visible in this architecture. Caches, registers, shadow registers, etc. will not be visible to even assembly language programmers. This uniformity allows a simplicity of structure in the implementation which will simplify the use of techniques such as microcode or programmed logic arrays (PLA's) for implementation of this architecture.

A third major concept is that of virtualization of address space. An address space will consist of a set of objects. These objects will be represented in this address space by their capability set. A locality set of objects will be realized in an intermediate access memory. A larger locality set will be represented in a self-managing secondary memory (SMSM) [Lipovski, et al, 1976] for objects. There will logically be an SMSM per process. Further virtualization will be supported by associative disks such as CASSM [Su and Lipovski, 1975], and finally by sophisticated but relatively standard data management systems on conventional very large scale mass storage. All of this virtualization will be completely transparent to the programmer.

It can be seen that these ideas lend themselves very readily to either a large scale single processor or to a distributed network of small systems. A small system might consist of a processor with one or more SMSM's and an associative disk.   The local system would draw its more extended locality sets from a central object repository driven by a very large machine.

3.0 PROCESSOR-MEMORY ARCHITECTURE

This section represents a progression of resolution of detail pointing towards an implementation.   In this section we discuss previous related architectures, the representation of capabilities, the representation of objects, the representation of localities and an abstract machine operation set.

3.1 Previous Architectural Proposals And Related System Issues

Several existing machine architectures have had substantial influence on our design; the most significant of these is the SWARD architecture [Myers and Buckingham, 1980; Myers, 1978]. The SWARD is a tagged, capability-based address architecture.   It has variable width and structured primitive data types, but does not directly support extended type objects. The SWARD design is motivated by the software reliability issues of protection, sharing and reliability. These were important considerations in our design, and it is not surprising that our design has very similar data formats.   The efficient support of objects, especially dynamic objects and objects representing relations and extended types is an important goal of our architecture and is responsible for major differences between our architecture and the SWARD architecture. Another architecture which influenced our design was the Plessey 250 [England, 1974], a capability-based system which implements capabilities by extended segmentation registers.

The INTEL 432 is an extended version of the capability register architecture implemented in modern technology.   The IBM System 38 implements capabilities by direct mapping of names to a very large address space.

Capabilities are important in our design. Their advantages are discussed in [Fabry, 1974] and [Feustal, 1973]. Implementation issues concerning capabilities may be found in [Fabry, 1974] and [Jones, 1980].   The use of capabilities as a protection mechanism is thoroughly covered in [Cohen and Jefferson, 1975] and [Wulf, 1974], which examines the HYDRA operating system.

The efficient implementation of capability-based architectures that must support large numbers of extended-type objects is discussed by [Lanciaux et. al., 1978] and [Gehringer,1979]. Our solutions are similar to theirs.

Extended-type objects are used to implement abstract data types. Abstract data types and control abstractions, which led to our idea for selectors, are discussed in [Liskov et. al., 1977].

Garbage collection will be important in our system. We anticipate that automatic garbage collection will be used at several levels within the memory system. Internal memory will probably be garbage collected using one of several methods which have been presented in the literature [Dijkstra, 1978], [Steele, 1976], [Baker, 1978] or [Deutsch and Bobrow, 1976].

3.2 Representation Of Capabilities

Fabry [1974] demonstrates that capability-based addressing overcomes the inadequacies of other addressing schemes in providing addresses for shared objects. Therefore an architecture to efficiently support object-oriented systems should use capabilities as addresses. The implementation of capabilities for such an architecture involves several design decisions.

1. the representation of capabilities

2. the representation of primitive data objects

3. the representation of extended type objects

4. the representation of relations between objects

Capabilities provide uniform and general naming, object addresses independent of execution context, manipulation of access control by non-system programs, and reasonable implementation of small protection domains. The representation of capabilities in a system should support these uses. One possible representation of a capability would be:

```
------------------------------------
| protection info | object label   |
------------------------------------
```

The protection information might be absent in an unprotected system or more likely might contain bits indicating the permitted kinds of access to the labeled object. Although the object label could be a physical address of the object, as in the Chicago Magic Number Computer, this requires the system to update all capabilities for an object whenever the object moves. A better solution introduces indirection as was done in the Plessey System 250. The object label is used to select an entry from the global object table (GOT). This entry contains the physical address of the object. Because the physical address is contained in only one place - the GOT - it is easy for the system to keep an object's address up to date. This is the first step is separating physical addresses from unique identifiers. To maintain the integrity of capabilities, object labels must never be reused for different objects; they must be unique identifiers for the object (this requirement is usually weakened, allowing unique ids to be reused after a long enough period of time has elapsed). At this point the representation of a capability would be:

```
-------------------------------------------
| protection info | unique id             |
-------------------------------------------
```

Unfortunately, this representation leads to inefficiencies due to the large amount of information that must be stored for each object. The capabilities for an object require enough bits so that unique ids don't have to be repeated too frequently, and the GOT entry for an object has to be large enough to contain the object's physical address along with other information about the object. There might easily be over 100 bits of overhead for an object in this kind of system. This overhead is not acceptable if there are a large number of small objects in the system.

Small objects are often components of large objects. For example, an integer (a very small object) is commonly a field within some larger object. (For example an object called Index might contain all of the indices for the code in a local execution environment.) It may be a field within an extended type object or a value cell within the activation record (also an object) for some procedure. This leads to another representation for capabilities:

```
----------------------------------------------------
| protection info | unique id | component offset|
----------------------------------------------------
```

This latest representation allows the system to treat all objects as either autonomous (an external object) or as a fixed component of an autonomous object (an internal object). Lanciaux, et al estimate that up to two-thirds of all objects in a system may be internal objects. An advantage of this view of objects is a smaller (and hence faster) GOT because internal objects all share their external object's entry. This, scheme also relieves the resource managers from handling internal objects individually (e.g., during garbage collection or swapping). Notice that this representation combines the unique id (indirect) kind of object label with the physical address (direct) kind of object label without the previous disadvantages of physical addresses in capabilities. Internal objects are assumed to be fixed, they have a static size and offset and their lifetime is no longer than the external object they are contained in. These strong requirements are necessary to avoid the disadvantages of physical addresses. Although most component objects will satisfy these requirements, not all will. External objects will contain (see discussion on representation of objects) capabilities for components which cannot be internal objects.

Rather than provide a whole capability for each internal object, one capability could be associated with many different offsets. Thus, an object with many internal components might only have one capability associated with it. The offsets into the object to select components will be determined by selectors. The selector associated with a capability always points to the current component of an object. Selectors are used in conjunction with capabilities to address structured objects. Selectors can be operated on to access different components of the same object at different times. Selectors will also be used as the addressing mode for unstructured objects. (See Section 3.4 for the discussion of selectors.)

With the introduction of selectors, capabilities now have a single, simple representation without an offset field.

```
 _ _ _  ---------------------------------------
| tag  |  protection info | unique id          |
 _ _ _  ---------------------------------------
```

This form restores complete separation of capabilities and physical addresses. Access to all objects will now require both a capability and a selector. Selectors, however, must be realized only in the context of a local execution environment.

The presence of the tag field in capability representations will depend upon the representation of extended type objects and how this representation handles embedded capabilities. To summarize, the advantages of this representation are:

1. Physical addresses for objects are limited to the GOT selectors and programmer invisible processor registers. registers.

2. Local internal objects (both small and large) may be accessed without going through GOT.

3. Local internal objects (both small and large) have low memory overheads, solving the small object problem.

4. The internal object concept improves memory management. (When one object is swapped in so are its static components.)

5. The GOT is smaller and faster.

Schemes similar to this one have been proposed by Lanciaux, et al [1978] and Myers [1978].

The chosen representation of capabilities allows neither capabilities nor instructions to be of a fixed width. A natural generalization of this idea allows

primitive data objects to be of arbitrary size. For example, an integer could have
the following format:

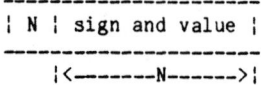

The representation of extended type objects is related to the proposed
representation of capabilities.   The concept of internal objects, introduced to
improve the efficiency of capability-based addressing of small objects, requires the
representation of extended type objects to contain both data (i.e., internal
objects) and capabilities (i.e., components of the object which cannot be internal
objects). Several alternatives exist that satisfy this requirement.

The Plessey System 250 and the Chicago Magic Number Computer support two kinds
of objects: (1) objects containing only data and (2) objects containing only
capabilities. Jones [1980] points out that this approach leads to an undesirable
proliferation of objects, many of which are used only to bind together one object
holding data and one holding capabilities. A better object representation would
allow capabilities and data to be mixed in the same object.

There are at least three ways to do this:

1. tagged capabilities

2. partitioned capabilities, and

3. fenced capabilities.

Tagging is used in the Rice machine and the B1700. By using tags, capabilities are
freely located throughout objects in these machines, but all memory words are tagged
and words identified as capabilities by their tags are protected from being
manipulated as ordinary data.

The HYDRA system uses software implemented partitioned capabilities. A system
of this kind partitions the data and the capabilities of an object into two separate
vectors which together represent the object. Only operations of the appropriate
type can be applied to the distinct vectors. Finally STAROS [Jones, 1980] uses the
fence method. All capabilities of an object are located at the front of an object
and a "fence" is used to separate them from the data portion of the object. Only a
simple bounds check is needed to determine if operations are acting on the correct
region of the object.

All three schemes seem to be viable; however, the first, tagged capabilities
seem to have an edge over the other two. Partitioned capabilities require two
separately managed memory segments per object, a minor disadvantage.   Dynamically
adding a capability to an object is much easier if the object's capabilities are
tagged rather than fenced. Tagged capabilities are first class primitives of the
system: they may be passed as parameters, returned as results, transmitted over
data lines, and loaded into registers. Tagged capabilities can be stored directly
in internal objects, even if they are structured, resulting in a more natural
representation of objects containing internal objects.

Finally, tagged capabilities can be a part of a tagged memory system which
might aid in the implementation of variable width primitive data objects. For these
reasons the representation of objects should use tagged capabilities.

3.3 Definition Of Execution Environments

The execution environment of a process and the name scope of a procedure are similar concepts. They determine a collection of objects which may be accessed by a set of instructions. All objects within a capability system are named by capabilities, so an execution environment or name scope is just a set of capabilities.

Because capabilities are relatively large compared to a typical span of local addresses, the cost of instructions will be high if capabilities are used directly within instructions. Execution of any instruction will always be done with respect to some execution environment, so it is practical to provide short names, called nicknames, for use by instructions in accessing objects.

All activations of a procedure (or process) have an associated execution environment; each execution environment is a set of capabilities and selectors stored within an array called a nickname table. The nickname used by a procedure's instruction to access an object is the nickname table index of the object's capability selector pair. If a procedure accesses less than 256 different objects, a nickname could be stored in 8 bit fields in the procedure's instructions. A variable length field, like the size field of object tags, could be used for nicknames in instructions.

The objects for which a procedure's nickname table holds capabilities would include: the procedure's code, the nickname table of the procedure's caller, or for some set of nicknames within the nickname table of its caller, data objects local to the procedure's activation, and arguments. The nickname table also contains selectors. A selector is used by instructions to access components of structured objects. For example, an array selector might simply contain the index of the desired component. A selector and a capability for a structured object together determine a component value of the structured object. One selector that is always contained in a nickname table is the instruction selector (the program counter). The instruction selector and the capability for a procedure's code object together determine the instruction to be executed next. The semantics of selectors will be discussed in more detail in the next section.

Since an activation of a procedure or process that may be running or suspended is determined by a nickname table, capabilities for nickname tables give access rights for the manipulation of other procedures or processes. Functions passed as arguments and returned as values, co-routines exception handling and other very generalized control regimes may be easily implemented by the use of capabilities for nickname tables. A procedure P's nickname table will usually contain a capability for the nickname table of P's caller, and it is this capability that is used by the return instruction to return to the caller.

Because procedures use nickname tables to access all data, the capabilities present in a procedure's table determine its entire address space. Capabilities for all objects accessed by the procedure that have a "lifetime" which extends over all invocations of the procedure may be stored in the nickname table once and left there for the life of the nickname table.

An object which a procedure has access to only during a particular invocation must have its capabilities filled in at invocation time. Typically, a nickname table would be created and then initialized with capabilities for objects which are "own" or "static" objects of the procedure. Before calling a procedure each time, the caller would load the called procedure's nickname table with capabilities for arguments to the procedure. After the procedure returns from the call, a set of capabilities within the nickname table would be used by the caller to access the returned results of the procedure.

3.4 Selectors And Locality Space Addressing

Selectors are objects of primitive type. Literal constants are embedded in instructions, but all other objects within a procedure's execution environment have an associated capability and at least one selector stored within the current nickname table. Instructions contain indices into the nickname table. The selectors in the nickname table determine the instruction operands. Although some instructions will operate directly on selectors, most instructions will use selectors to access components of structured objects or entire objects.

To access the components of an array object, a selector for the array would be initialized and then employed by instructions to access some component. Subsequent accesses of additional components would be accomplished by manipulation of the selector before each access. For example, the following loop

```
        for i := 1 to 10 by 2 A[i]:= 0
might be translated (roughly) to
        S <- create_selector (A, from_to_by)
        S <- (1,10,2)
Loop:   S@ <- 0
        next S, loop
```

Here, "next S" is an example of an instruction that manipulates selectors, while "s@ <- 0" is the move instruction using a selector to access a component of "A". Observe that an object's selector has two parts: a mode and a value. The mode determines the pattern of access. A common mode for accessing arrays would be the "from_to_by" mode presented in the example. A small number of built-in modes will be provided by the hardware for the standard types. If the access pattern for a data structure does not correspond to a predefined mode, then a procedure implementing such an access pattern on the data structure's selector may be written. The selector's node now contains a nickname for this procedure. Selectors form a convenient control abstraction mechanism, much like the iterators of CLU [Liskov. 1977] or the access maps of BLISS [Wulf, et al, 1971].

We anticipate many objects to be compositions of a number of structured primitive data types. For example, a symbol table might be an array of records. One selector would be used to access a particular array element (a record) and a second different selector to access a field within the selected record. In such a case the first selector would be associated with the object's capability and the second selector with the first selector. These associations are maintained by a field within each selector with the nickname of the selector or capability that is their "child".

One right that is controlled by a capability's access rights field is the right to create certain kinds of selectors for the object specified by the capability. When an object, and in particular an extended type object, is created the system creates a capability for it with all rights enabled. Before passing such a capability out of the type manager for the object (where creation takes place) the type manager disables the right to create component selectors for this object. Now, other users of the object cannot "open" the object through the use of selectors, and may only pass the object whole to other procedures. Type managers are able to use selectors on objects of their type because the necessary selectors were created when the manager created its first object. These selectors are maintained within the type managers execution environment and could be passed out to trusted procedures.

In addition to the "from_to_by" mode used in the example above, other primitive modes would be provided by the system. "All" is a mode that gives access to the entire object. Selectors of this mode can always be created from the capability for the object. "Sub" and "dot" are restricted selectors like "from_to_by". They may only be created if the capability for the object contains the right to create component selectors, that is selectors which access components of a structured object. "Sub" accesses a component of an array with a specific subscript. "Dot" accesses a specific field of a record structure.

Typical operations on "sub" selectors might be:
  initialize - select a specific component
  next - select next component
  previous - select previous component
"Dot" selectors might have only "initialize" as an operation, choosing a specific field. "From to by" selectors may be used on arrays or records for stepping through the components. This selector may also be used to step through linked lists. For example this code

```
Ptr <- create_selector (A, "from_to_by")
Ptr <- (P, Q, 2)
```

might initialize a "from_to_by" selector to access a linked structure in area A.   P points to the first node in the list to be accessed, Q the last, and the 2 indicates that pointers to successive nodes are obtained by following a link field stored as the second field of each node.

In addition to the primitive selector modes, user-defined modes may be created. In this case, the selector contains a nickname of a user-defined procedure which computes selector values.

Efficient access to almost any data will depend on the efficiency with which selectors select components of objects. It is anticipated that the hardware will provide ancillary registers, invisible at the machine language level, that are automatically maintained during the use of selectors for the predefined modes. These supplementary registers could improve efficiency by transforming address calculations for predefined access patterns into more efficient ones (for example multiplicative matrix index calculations into additive index calculations), by introducing parallelism through calculation of the next access address before it is required, and by the elimination of indirection through the maintenance of physical addresses rather than logical or capability addresses.

3.5 Representation Of Primitive Data Objects

Data objects are either of primitive type, that is a native machine type, or extended type. An extended type object is defined in terms of component objects each of which is of extended type or primitive type. The representation of extended type objects has already been mentioned briefly. The primitive types must be varied and flexible enough to provide convenient elements from which all other types are built. This means that in addition to the standard computational modes like integer, real, and Boolean, modes should be provided for structuring data into arrays, records and strings.

The greatest flexibility is achieved when the size and precision of the primitive data types are not bound to a fixed size word. The SYMBOL [Rice and Smith, 1971], B1700 [Wilner, 1972], and SWARD [Myers, 1978] machines all use storage cells that may be of arbitrary size for holding values of the primitive data objects. In SWARD all storage cells are contiguous sequences of an integral number of 4-bit tokens. This seems to be a good compromise between fine enough granularity to prevent wasted space within storage cells and saving bits in address or length fields. For example the proposed format for integer data is:

```
---------------------------------------------
| int  | size |        value               |
| code |      |                            |
---------------------------------------------
     | 1 + size 4-bit units   |
```

The first two fields constitute the "tag" for this primitive data object. Note that this tag contains a size field which specifies the length of the field used for holding the integer's value. During the entire lifetime of a data object, its tag, which includes the size information is fixed. The size of a primitive data object's storage cell does not change during the lifetime of the data object. This is important because it is a serious restriction but a necessary one. Memory management would be far too difficult if each primitive data object could change the width of its storage cell during a program's execution. This kind of dynamism may be supported, but at a higher level, perhaps through the allocation of new larger or smaller storage cells and the discarding of old cells.

Tagged architectures have advantages that are well known (see [Myers, 1978] and [Feustal, 1973]) and that have been taken advantage of in several machines. The advantage of tagging capabilities has already been discussed, and it seems natural to extend tagging to cover all primitive data types. The storage cell for all objects has two parts: a fixed descriptor part, the tag, and a fixed-width varying-content part, the value. The tag will frequently be composed of two parts, as in the integer example, a code identifying the type of the object and a size field containing size information about the value part.

A data object's tag should be short to keep the overhead of tagging to a minimum. The code part of an objects' tag may be kept in a single 4-bit unit. An escape code (for example 0000 or 1111) could be used to indicate that one or more subsequent units be used for the objects code part. This allows the architecture to support more than 16 primitive data types.

Since primitive data objects can range from Booleans to large arrays, the size part of the tag must be large enough to reflect this range. However, it is desirable to have as small a size part as possible to minimize the overhead of objects' tags. By encoding the size in a variable number (fixed at object creation time) of units, small objects can have small size fields within their tag and large objects can have large size fields within theirs. The following formats for the size part of tags are suggested:

size	storage units	format
0-3	1	¦ 00xx ¦
0-63	2	¦ 01 xx,xxxx¦
0-16K	4	¦ 10 xx xxxx xxxx xxxx ¦
0-1G($10^{**}9$)	8	¦ 11 xx xxxx xxxx xxxx xxxx xxxx xxxx xxxx¦

Using these formats sizes from 0 to 1,073,741,823 may be represented while using a field that is either 1,2,4 or 8 units long. Examples:

short short integer	10/x	-7..7
short integer	11/xx	-127..127
medium integer	13/xxxx	-32767..32767
long integer	14A/xxxxxxxxx	-10[12]..10[12]
(slashes separate tag field from value field)		

Many conventional machines have no way to represent undefined values (typically an integer stored in n bits may represent $2^{**}n$ distinct values). This makes detection of certain run-time errors much more expensive. Several new machine architectures extend the basic types with the value "undefined" in order to aid run-time error detection [Myers, 1978] and [Tannenbaum, 1978]. It has been proposed by some [Feustal, 1973] that in a tagged architecture the tag could contain a bit indicating that the objects value is undefined. This method has been rejected for two reasons:

(1) It is inconsistent with the view developed so far that the tag is a <u>fixed</u> descriptor for an object. (2) Structured types may have one tag for many values. Instead, I suggest that the range of values for the primitive types include a representation of "undefined" where appropriate.

```
                              |
              -------------------
Integers:     |1| size| value  |
              -------------------
```

Value is a 2's complement representation of the integer. Undefined is represented by 800...0.

Examples:
1:0/5	5
1:0/8	undefined
1:0/F	−1
1:3/FFFF	−1
1:3/7FFF	32767
1:48/00000001	1
1:48/80000000	undefined
1:C0000000/5	5

Integers in the range −32K..32K require a storage cell 24 bits wide (see 4th and 5th examples) compared to 16 bits required on many machines. Twelve bits of overhead are required for all integer objects requiring a range larger than this but no larger than − 4,611,686,018,427,387,903.   4,611,686,018,427,387,903.

```
                    12
                    |
          -------------------------------------------
Reals:    | 2 | size  | signs | exp | mantissa |
          -------------------------------------------
           1   1..8      1       2    3*(size+1)
```

signs   − sign bits of exponent and mantissa
exp   − 0..256, base eight exponent
mantissa − octal mantissa, 4*(size +1) octal digits (3*(size +1) memory units) long.

Undefined is represented by a 1xxx signs field. The approximate range of reals that can be represented is $-(10^{**}230)..(10^{**}230)$. The precision varies with the size of the mantissa field.

Examples:
2:0/0:01:100	1.0
2:1/0:01:100000	1.0000
2.1/1:01:100000	−1.0000
2:1/2:01:100000	0.1000
2:1/3:01:100000	−0.1000
2:3/0:0C:112402762000	
2:3/0:01:311037552421	
2:2/0:00:000000000	0.0
2:2/8:05:123456712	undefined

44 bits are required to give approximately the same precision as IBM 360 single precision reals of 32 bits. CDC 6600 single precision reals require 60 bits, in this format the same precision could be obtained with about 68 bits. CDC double precision requires 120 bits, however in this format the same precision would also cost about 120 bits.

```
                                  ¦
                       ----------------------
Booleans:             ¦ 3 ¦ size ¦ value ¦
                       ----------------------
                         1   1..8   ¦size/2 ¦
```

Value is a string of Booleans of length size where each Boolean in the string occupies two bits. Two bits are used rather than one in order to provide an undefined value in the type Boolean. False is represented as 00, true as 01, undefined as 1x.

Examples:

3:1/4	T
3:2/4	TF
3:2/6	TU
3:50/00000000	FFFFFFFFFFFFFFFF
3:0/	<empty>
3:44/19	FTUT

```
                                  ¦
                       ----------------------
Character String:     ¦ 4 ¦ size ¦ value ¦
                       ----------------------
                         1   1..8   2*size
```

Value is a sequence of 8-bit characters, size characters long. One character is reserved for undefined, perhaps FF.

```
                         ¦
                    ----------
Void:              ¦5¦ size ¦
                    ----------
```

Void: This type is used primarily in conjunction with the flex type. The representation of a void object has no value part. The only value in the type void is nil. [Should void also have undefined in its set of values?] A programming example of the use of void is the recursive definition of the type tree:

```
         tree  =record
                ¦key : integer;
                ¦lson, rson : one of (void, tree)
                end
```

```
                    ¦
             ----------------------
array : ¦tag-part ¦ value   ¦
             ----------------------
```

where the tag-part is :

```
----------------------------------     ---------------------------------
¦ 6 ¦ N ¦ bound-1 ¦ bound-2 ¦  ...  ¦ bound-N ¦ component tag¦
----------------------------------     ---------------------------------
  1 w:1:8   w            w                   w
```

N is the number of dimensions of the array. The bounds in the tag are the bounds of the dimensions of the array. The components of the array are stored without tags in the value field. A single tag applies to all component values, the component tag.

Examples:
6:1:A:1:3/0000:0000:0000: . . . :000
array[1..10]of -32767..32767

6:1:4:3:1/0:0:0:0
array [1..4] of Boolean

6:42:A0:5:1:2/000:000:. . . :000
array [1..160,1..5] of -4095. . 4095

6:1:A:6:1:4:4:1/00:00:. . . :00
array [1..10] of array [1..4] of char

```
                                          ¦
           -------------------   -------------------
Structure: ¦ 7 ¦ N ¦ tag-1 ¦ ... ¦ tag-N ¦ value ¦
           -------------------   -------------------
             1  1..8
```

p-5 N is the number of components of the structure.   Tag-j is the tag for the structure's jth component.

Examples:
7:2:5/1:1/xx:1:1/xx   record x: -127..127
                                        y: -127..127
                                 end

```
                      ¦
           ------------------------
flex:      ¦ 8 ¦ size ¦ value    ¦
           ------------------------
```

Size is the size of the value field. The value field is used to hold a tagged data element.   Instructions will interpret a flex object as if it were the object represented by the flex value field alone. Note that the tagged element held in the value field can have its tag and hence type changed at run-time.

```
                     ¦
           -------------------
nickname : ¦ 9 ¦ size ¦ value¦
           -------------------
             1   1..8   1+size
```

Value is the positive index of some entry in the current nickname table. The entry contains a selector or capability for some object. Nicknames are locally valid addresses. Undefined is represented by a value field of zero.

```
                     ¦
           ----------------------------------
capability : ¦ A ¦ unique id ¦ access rights ¦
           ----------------------------------
             1      8
```

Capabilities are addresses that allow arbitrary external objects to be addressed.

The unique id and address rights fields have already been discussed in the section on representation of capabilities.

```
                 -------------------------------------
indirect I:      ¦ B ¦ nested tag for value ¦ nickname ¦
                 -------------------------------------
```

Indirect I allows a tag to be associated with a value that is not contiguous with the tag. For example: 11 :[1:1] :[9:1:xx] is treated as an 8 bit integer the value of which is located at address specified by nickname xx.

```
          ----------------------------------
          ¦ B ¦ tag for V ¦ nickname for V¦
          ----------------------------------
                           ¦
                           ------------------->-----
                                               ¦ V ¦
                                               -----

                     ¦
                     ----------------------------------
indirect II :        ¦ C ¦ nickname for value's address¦
                     ----------------------------------
```

References to cells of type indirect II causes the machine to reference the object pointed to by the address part of the Indirect II cell. Like Indirect I the automatic dereferencing is invisible to the machine's instructions. However, unlike Indirect I, indirect II points to a tagged value. (Indirect I pointed to an untagged value).

```
                     ¦
        --------------------------
code:   ¦ D ¦ size ¦ value      ¦
        --------------------------
                    size
```

Executable code is stored in an object of type code.
Extended Type Objects:

```
                      -------------------------------------------
                      ¦ E ¦ type-manager cap ¦ nickname   ¦
                      -------------------------------------------
```

Extended type objects (ETO's) have two parts: a capability for the type manager of the object and a nickname for a selector stored in the type managers nickname table which is used to access components of the ETO.

Area

```
        --------------------------
        ¦ F 0 ¦ size ¦ area     ¦
        --------------------------
          2      1..8
```

Ptr

```
        ¦
        -----------------------
        ¦ F1 ¦ size¦ value ¦
        -----------------------
          2     1..8
```

Selector

```
        ---------------------
        ¦S¦C¦   PTR         ¦
        ---------------------
         2 8
```

The S field gives the type of the selector, (address pointer or nickname table entry for a type manager). The C field points to a "child" entry in the local nickname table. The PTR field is either a pointer to a physical address or a nickname table entry.

4.0 IMPLEMENTATION CONSIDERATIONS

It is important to make a feasibility demonstration for the implementability of this proposed architecture. The concept basis set which is used here has all been realized at least partially in hardware or in software in the past. There are no totally untested concepts being used. Capabilities have been implemented in hardware in a number of machines. The Plessey 250 [England, 1974] and the Cambridge CAP [Needham and Wilkes, 1978] machine are examples. The INTEL 432 is a more recent example. These systems all combine capabilities and addresses at the architecture level through extended instruction visible segment registers. Capabilities have, of course, been implemented in software in a number of machines including the HYDRA [Cohen, et al., 1975] operating system for C.mmp and the CAL system [Lampson and Sturgis, 1976] at the University of California at Berkeley. It has also been shown that segment based architectures can also be converted to capability oriented architectures with small overhead [Hoch and Browne, 1980]. Object oriented architectures are relatively newer. The IBM System 38 realizes an object-oriented architecture through direct mapping of a very large linear address space. Abstract data type object concepts have been very thoroughly developed in the HYDRA operating system and in a number of data oriented higher level languages such as CLU and ALPHARD.

The locality concept is prominent in modern day architectures. Locality and virtualization are closely tied together. The early Burroughs architectures beginning with the 5500 dealt with localities of segments (a primitive definition of objects). Of course, virtualization has been implemented very widely in terms of fixed sized blocks (i.e., pages). A principal concept of this architecture is that a virtualization of objects should be extended even to very small objects and that this can be efficiently done. The principle of locality also appears in shadow registers, caches, etc. in many previous architectures.

Defining operations directly upon objects has been done in software (HYDRA, System 38).

A key point in this proposed architecture is the ability to deal with variable length processing. Variable width processing units have been defined and described by several authors. Reconfigurable architectures are emerging through the Texas Reconfigurable Array Computer (TRAC) project [Premkumar, et al., 1980], [Sejnowski, et al, 1980], [Kapur, et al., 1980] and other network based architectures.

A key element in the implementation of a total system is a requirement for a multi-level object oriented memory. A prototype of such a device is now being built on the TRAC project. The "associative disk" concept is common in the literature and has been implemented.

It is clear that the simplicity of the instruction set and the uniformity of the memory structure will make the construction of microcode and/or program logic arrays to support complex object operations relatively simple. It is this uniformity and the small number of concepts which must be implemented which makes the implementation of this system in hardware feasible.

5.0   REFERENCES

Baker, H. G., "List Processing in Real-time on a Serial Computer", CACM 21(4), 280-294 (1978).

Cohen, E. and Jefferson, D., "Protection in the HYDRA Operating Systems", Proc. 5th ACM Sym. on O. S. Principles: ACM Operating Systems Review 9(5)141-160 (November 1975).

Deutsch, L. P. and Bobrow, D. G., "Efficient, Incremental, Automatic Garbage Collector", CACM 19(9), 522-526 (1976).

Dijkstra, E. W., et.al., "On-Fly Garbage Collection - An Exercise in Cooperation", CACM 21(11), 966-975 (1978).

England, D., "Capability Concept Mechanism and Structure in System 250", IRIA, Int. Workshop Protection in Operating Systems, 63-82 (August 1974).

Fabry, R. S., "Capability-based Addressing", CACM 17, 7, 403-412 (July 1974).

Feustal, E. A., "On the Advantages of Tagged Architectures", IEEE Trans. on Computers, C-22, 7, 644-656 (July 1973).

Gehringer, E. F., "Variable Length Capabilities as a Solution to the Small-Object Problem", Proc. 7th Symp. on Operating Systems Principles, 131-142 (1979).

Gligor, V. D., "Architecture Implementations of Abstract Data Type Implementations", 6th Annual Symposium on Computer Architectures, 20-30 (1979).

Hoch, C. G. and Browne, J. C., "An Implementation of Capabilities on the PDP-11/45", Operating Systems Review, 14, July 1980, 22-32.

Jones, A., "Capability Architecture Revisited", Operating Systems Review 14(3), 33-35 (July 1980).

Kapur, R. N., "Premkumar, U. V. and Lipovski, G. J., "Organization of the TRAC Processor-Memory Subsystem", Proc. of the National Computer Conference, 1980, 623-629.

Lampson, B. W. and Sturgis, H., "Reflections on an Operating System Design", CACM 19(5), (May 1976).

Lanciaux, D., Schiller, C. and Wulf, W. A., Supporting Small Objects in a Capability System, CMU-CS-78-107, (1978).

Liskov, B., Snyder, A., Arhison, R. and Schaffert, C., "Abstraction Mechanism in CLU", CACM 20(8), (August 1977).

Myers, G. J., and Buckingham, B. R. S., "A Hardware Implementation of Capability-Based Addressing", Operating Systems Review 14(2), 13-25 (October 1980).

Myers, G. J., "Storage Concepts in a Software-reliability-directed Computer Architecture", Proc. 5th Annual Symp. on Computer Architecture, 107-113 (1978).

Premkumar, U. V, Kapur, R. N., Malek, M., Lipovski, G. J., Horne, P., "Design and Implementation of the Banyan Interconnection Network in TRAC", Proc. of National Computer Conference, 1980, 633-643.

Rice, R. and Smith, W. R., "SYMBOL - A Major Departure from the Classic Software-dominated von Neumann Computing Systems", Proc. SJCC 1971, 575-587.

Sejnowski, M. C, Upchurch, E. T., Kapur, R. N., Charlu, D. P. and Lipovski, G. J., "An Overview of the Texas Reconfigurable Array Computer", Proc. of National Computer Conference, 1980, 631-641.

Steele, G., "Multiprocessing Compactifying Garbage Collection", CACM 18(9), (September 1975); see also corrigendum in CACM 19(6), 354 (June).

Su, S. W. and Lipovski, G. J., "CASSM: A Cellular System for Very Large Data Bases", Proc. of International Conference on Very Large Data Bases, Farmington, MA, 456-472 (1975).

Tannenbaum, A. S., "Implications of Structured Prog. for Machine Architecture", CACM 21 (3), 237-246 (March 1978). "Collection", CACM 19(9), 491-500 (1976).

Wilner, W. T., "Burroughs B1700 Memory Utilization", Proc. of the Fall Joint Computer Conference, AFIPS, 1972, 579-586.

Wilner, W. T., "Design of the Burroughs B1700", Proc. of the 1972 Fall Joint Computer Conference, AFIPS, 1972, 489-499.

Wulf, W. A., et.al., "HYDRA: The Kernel of a Multiprocessor Operating System", CACM 17(6), 337-345 (June 1974).

Wulf, W. A., Russell, D. B. and Habermann, A. N., "A Language for Systems Programming", CACM 14 (12), 780-790 (1971).

Wulf, W. A., Levin, R. and Harbison, S. P., HYDRA/C.mmp: An Experimental Computer System (McGraw-Hill, Inc., New York, 1981).

SUPER FREEDOM SIMULATOR PAX

*   *
Tsutomu Hoshino, Tomonori Shirakawa,
**   *
Yoshio Oyanagi, Kiyo Takenouchi,

University of Tsukuba,
* Institute of Engineering Mechanics,
** Institute of Information Sciences,

Sakura-mura, Niihari-gun, Ibaraki-ken, JAPAN,

and

Toshio Kawai

Keio University,
Department of Physics

Hiyoshi, Kouhoku-ku, Yokohama-shi, JAPAN.

ABSTRACT

A parallel array computer PAX has been developed for the use in scientific calculations. The machine is an MIMD, nearest neighbor connected processor array with special synchronization hardware. Several typical applications have been tested; among them are Monte Carlo simulations in theoretical physics, and the fluid dynamics solving the Navier-Stokes equations. Very high efficiency has been attained in the parallel processing of these problems, that can be attributable to the well-matched architecture with the inherent parallelism in the nature. The machine can be a prototype for SUPER FREEDOM SIMULATOR, a future VLSI parallel computer with huge number of processors.

Keywords. parallel computer, multiprocessor, processor array, parallel processing, parallelism, VLSI computer, super computer, simulator, MIMD, nearest neighbor connection, Monte Carlo simulation, Ising model, Heisenberg model, lattice gauge theory, fluid dynamics, Navier-Stokes equation.

1. Introduction

The nineteen-eighties will be the era of revolutionary change in the computer design. Present VLSI technology with its rapid growthrate will make it a reality that various types of dedicated computer systems are technically and economically feasible.

Super-computer is certainly one of these dedicated computers, specially tuned to the scientific numerical calculations. Actually the needs of scientific calculation is immense. Computational speed of 1000 times faster than the presently available is required.

The scientific calculation is characterized by the universal principle of physics : "action through medium," stating that the physical action comes from the immediate neighbors of the medium or field. The principle assures that, if it is well-utilized in the computer design, the interprocessor communication and the calculation can be limited in the neighbor and inside of each processor, respectively. The multiprocessor architecture that utilizes this inherent property of the proximity is well-known nearest neighbor mesh (NNM) connection of the processors. Straightforward way of implementing the physical problem on the NNM processor array is such that the physical space is directly projected onto the processor array: we call this "direct mapping."

Thus the NNM array with direct mapping makes the parallel processing go autonomously, with the least intervention from the other processors. This leads to a high efficiency of parallelism. The autonomous parallelism is a close analog to the physical nature itself. In this sense, the computation is "simulating" the nature.

Another salient property of the nature is its infinitely large degrees of freedom. For example, the material consists of the infinitely large number of molecules, atoms, and the elementary particles. These particles move uniformly as a single rigid matter, or randomly as if each has independent freedom to move. The latter case appears in the simulations of gauge field theories where dynamical variables are allocated on each link (or bond) of a hypercubic lattice, which interacts with the neighbors.

The simulation of these infinitely large degrees of freedom requires that the processing units be large in number and each unit can execute independent parallel tasks asynchronously. This implies, in terms of the computer architecture, the processor array must be of MIMD type.

Summarizing the several requirements for scientific calculations, we need a special-purpose computer system, such as ;

1. Array of processing units(PUs), where
2. Each PU is identically designed.
3. Super-parallelism, i.e. number of PUs can be very large; as large as we can implement.

4.   Only nearest neighbor PUs are connected.

5.   MIMD structure, i.e. each PU can execute its own program.

We introduce here an experimental machine PAX (Processor Array eXperiment. It is the authors' wish that the machine will be used only for the peaceful purposes.) The objective of the development is the feasibility study of future super-parallel machine; what we call SUPER FREEDOM SIMULATOR.

2.   Pilot Machine PAX-32

The presently installed PAX-32 has been operating since 1980 in numbers of applications. The details will be reported elsewhere [1]. A brief summary is introduced here. System PAX-32 consists of 32 processing units(PUs). Each unit is essentially a microcomputer with 2 CPUs (MC6800 and AM9511), and several memories. All 32 PUs constitute 8 X 4, 2-dimensional array with the nearest neighbor connection, as shown in Figs. 1 and 2. The NNM connection is made by sharing the memory with nearby PUs. There is a control unit (CU), a microcomputer with MC6800, that takes care of the program and data loading, data retrieval and the synchronization of PUs. Through CU, the PU array is connected with the HOST minicomputer (TI990). The array acts as a number cruncher, and is controlled by the HOST's main program through calling subroutines. The parallel processing carried out by PU array is described with a PASCAL-like high level language SPLM. The language provides, other than the conventional statements, the data transfer with the nearest neighbor processors. There are several supporting softwares for parallel processing; among them are synchronization procedure, broadcast of data to all PUs, etc.

The applications have so far been made in several typical examples. For problems that keeps the proximity of the process, the linear scaling law was established, i.e. the performance is proportional to the number of PUs. Those problems without the proximity, such as the molecular dynamics, still follow this linear scaling law, since the overhead caused by the global data exchange is fairly small compared with the net calculation carried out by each PU.

The application problems here introduced are Monte Carlo simulation of spin and gauge systems with various interactions, and the fluid dynamics governed by the Navier-Stokes equation. These problems are well fitted-in with PAX architecture, since these posses the proximity of physical actions in their nearby space points.

3.   Ising and Heisenberg Models Implemented on PAX

The elements such as Fe, Ni, and Co have the magnetic moment due to the electron spin. When the material becomes ferromagnetic, the spins are aligned in the same

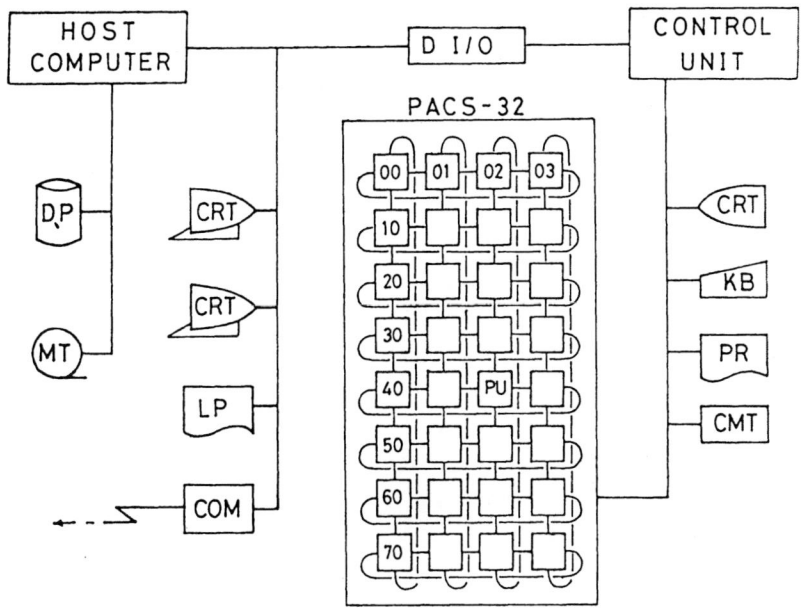

Fig. 1.   System configuration of PAX-32.

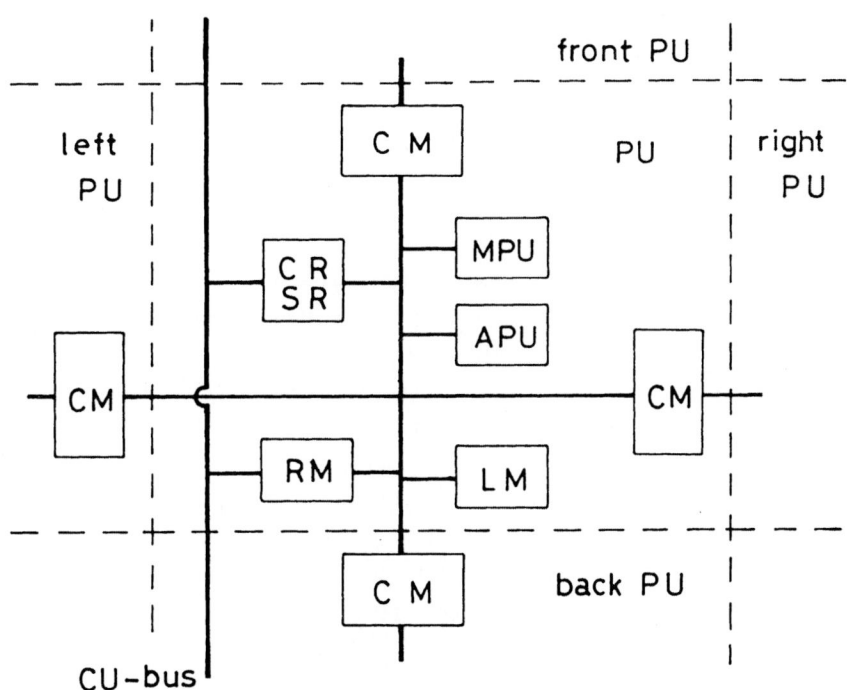

Fig. 2.   Configuration of a Processing Unit.

direction. The electron acts as a small magnet. A unit vector is used to indicate the direction of an electron spin.

The model of magnetic material is thus considered as a set of spins allocated on the lattice sites. Ising model is a model that restricts the direction of spin in two-directions: upward and downward. Heisenberg model is not subject to such a restriction; the spin vector is three-dimensional.

The ferromagnetic material has the strong interaction between the neighboring spins that favors the parallel spins in any direction. Thermal fluctuation proportional to the temperature, on the other hand, tends to disorder the spins. We consider here a spin system on a two-dimensional square lattice with nearest neighbor interaction. A spin has an interaction with the nearest four spins and does not receive any direct influence from the rest of the system.

Though the direct interaction comes from the immediate neighbor, the influence propagates beyond there. The correlation is detected between the spins situated in far distance. The average distance weighted by the correlation is called the correlation length. For the material in very high temperature, the correlation length is almost zero. As the temperature gets close to the Curie temperature, the correlation length becomes very large. Below the Curie temperature, the material shows spontaneous magnetization.

The computer model simulates this process following the heat bath method, such that the spin is randomly set according to the rule: probability of the upward spin setting is proportional to $\exp(\beta \sum s_\mu)$, and that of the downward setting to $\exp(-\beta \sum s_\mu)$, where $\sum s_\mu$ is the sum of the nearest four spins, and β is the inverse temperature. The correlation between spins is also obtained, as a function of the distance between the spins, such that $c_k = \sum s_{ij} \, s_{ij+k}$. The model has the proximity, and the high efficency in parallel processing can be attained.

Two-dimensional Ising model allocates the spins on the plane lattice with the cyclic boundary condition. The physical space is directly mapped onto the PU array of PAX-32. For example, 16 X 16 spin lattice is divided into 8 X 4 sub-domains. Each sub-domain includes 2 X 4 spins, and is taken care of by single PU. The program consists of the initial spin setting followed by the iteration loop that includes the spin update and the correlation calculation. In each iteration the direction of a spin is so determined that the new spin direction tends to be correlated with the sum of the directions of nearest neighbor spins. The strength of the alignment is inversely proportional to the temperature.

The efficiency of parallel processing is defined as the ratio of the time for net calculation over the total computation time. The ratio is expressed by 1 – (overhead associated with the parallel processing). Generally there are two causes of the overhead: one is due to the idling of processors that is the direct consequence of the uneven work load among the processors. However, this is not the present case, since the work load is identical if we allocate the same number of spins to each PU. The other cause of overhead is communication between the adjacent PUs. This is interpreted as an overhead, in a sense that the interprocessor communication does not appear in the conventional computers with single processor. This part can be measured on PAX-32, together with the measurement of computation times for the spin update and correlation of spins. These are shown in Table 1 with the efficiency of parallel processing. The efficency attained is quite high. The extension of the performance obtained for this model needs more carefull study of the computational time as a function of the problem size and the PU array size as well. This will be made for more realistic Heisenberg model described below.

The Heisenberg model expresses three-dimensional spin vectors allocated in the two-dimensional square lattice. The treatment of the model by PAX computer is similar to that of the Ising model. The computational time measured for each part of the processing is shown in Table 2. The performance is expressed in terms of the efficiency which is a percentage of the processors effectively working.

Table 1. Performance obtained in the simulation of 2-dimensional Ising model.

Number of Spins	Computing Times per Iteration			Efficiency (%)
	Spin Update (s)	Correlation (s)	Communication (s)	
16x16	0.0485	0.0965	0.0350	75.8
32x32	0.173	0.488	0.0635	90.4
64x64	0.650	3.38	0.121	97.0

Table 2. Performance obtained in the simulation of 2-dimensional Heisenberg model.

Number of Spins	Computing Times per Iteration			Efficiency (%)
	Spin Update (s)	Correlation (s)	Communication (s)	
16x16	0.1514	0.4819	0.1603	74.7
32x32	0.5673	1.727	0.2315	89.9
40x40	0.8752	2.535	0.2443	92.8
64x64	2.155	12.10	0.5093	96.4

The efficiency measured can be summarized as a scaling law:

$$\text{Efficiency} = 1 - \frac{1.82(n_1/m_1 + n_2/m_2) + 0.47\ d\ (n_1/m_1 + 23.9)}{17.8\ N/M + 2.72\ d\ (n_1/m_1 + N/M + 4.66)}$$

where the number of spins is denoted by $N = n_1 \times n_2$, the number of processor array by $M = m_1 \times m_2$, and the maximum distance to measure the correlation by d. The second term of the right-side of the efficiency is the overhead due to the interprocessor communication. If we extend the performance to the PAX with 4096 PUs, allocated in 64 X 64 square array, the problem with the spins of 64 X 64 (therefore 1 spin for 1 PU) with 32 maximum correlation distance will be processed with 36.7%. The performance is degraded due to the fact that the data routing for correlation (i.e. term 0.47d X 23.9) takes major part of the communication overhead, while the net calculation for correlation itself (term 2.72 d N/M) is relatively small. However, this is the most pessimistic case, that the user will not consider.

The estimate of the efficiency was made for more realistic case, i.e. 1024 X 1024 spins with maximum correlation distance 512 is simulated. The performance for this case is estimated as 97.5%, that is quite high. This is again due to the communication overhead, that becomes fairly small: communication overhead $0.47\ d\ (n_1/m_1 + 23.9)$ vs. net calculation of correlation $2.72\ d\ N/M$. This is because the communication takes time proportional to the number of spins, while the correlation to the product of correlation distance with number of spins. Thus we can summarize the performance of PAX with 4096 PUs for this Heisenberg Monte Carlo model, such that about 3990 PUs will work effectively, or 3990 times speed up can be expected.

4. Lattice Gauge Theory Implemented on PAX

Quantum Chromodynamics, QCD, a unique candidate to be a theory of strong interaction in elementary particle physics, could give qualitative results for the short range properties of hadrons (proton, neutron, pi-meson, etc.) in terms of perturbation expansions. On the other hand, at long distances (of order of 10E-13 cm) the perturbation method fails due to divergences, so that nothing could be predicted on the long range properties such as the mass spectrum of hadrons. K. Wilson [2] provided a non-perturbative approach, called lattice gauge theory, by formulating gauge theories on a four-dimensional hypercubic lattice. One can analyse the theory numerically by a Monte Carlo technique, using the equivalence between euclidean quantum theory and statistical mechanics.

The system consists of variables U_{ij} (an element of a group G), which is associated with the link joining a pair of nearest neighbor sites i and j. The action describing the system is

$$S = \sum T_r (1 - U_{ij} \, U_{jk} \, U_{k\ell} \, U_{\ell i})$$

where i, j, k, and ℓ circulate around an elementary square of the lattice. The partition function at temperature $T = 1/\beta$ is

$$Z = \sum \exp(-\beta S)$$

where the sum runs over all possible values of the link variables U_{ij}. The Monte Carlo technique consists of setting up a Markovian process which simulates an ensemble of states in the thermal equilibrium. In the last years many results about the phases of lattice gauge theories have been obtained. Lattice gauge theory provides the most promising approach toward the strong interaction physics.

To be concrete, let us realize the lattice system of size 4 X 4 X 4 X 8 along x, y, z and t directions, respectively, on the present PAX-32. We choose U(1) for the gauge group G, i.e. each link variable U_{ij} is a complex number of modulus 1, $U_{ij} = \exp(i \Theta_{ij})$, where Θ_{ij} is a real parameter. The system is mapped on the processor array in such a way that each PU performs calculations of the 4 X 4 sub-lattice which belongs to a fixed z and t coordinates. Since four links are associated with one lattice site, the number of link variables allocated to a PU is 4 X 4 X 4 = 64. The boundary condition is a cyclic one, which is best suited to the PAX architecture. The Monte Carlo methods used here are heat bath algorithms. At the beginning of each iteration, a PU communicates with the neighboring PUs to exchange the data for the link variables. During one iteration step, a PU performs serially Monte Carlo calculations on the links allocated to it without any communications with other PUs. The ratio of inter-PU communication to inner-PU calculation is estimated to be less than 1%, if implemented on PAX-32. One iteration will take 3 sec. The calculation of 1 X 1, 1 X 2 and 2 X 2 Wilson loops will take 1.5 sec. If we extend the performance to the PAX with 4096 PUs, connected in 16 X 16 X 16 cubic array, the problems with 16 X 16 X 16 X 16 lattice gauge model will be solved. The overhead of inter-PU communication for a Monte Carlo iteration will be less than 0.5% depending on the gauge group.

5. Fluid Dynamics Problem Implemented on PAX

Fluid dynamics problem has been a challenge for computational physicists. This three-dimensional, non-linear problem can be best simulated by tracing the evolution of velocity and density as functions of space. This method, called the primitive method, however suffered from vast computer time and stability problem. The discovery of a stable explicit algorithm, Rational Runge Kutta method (RRK) [3],[4], made the primitive method practical, if the computer speed is greatly improved by the super simulator such as PAX. The program for two-dimensional viscous compressible flow in a rectangular region was written for PAX, using the RRK algorithm. The program analysis reveals the high efficiency of parallel computation for continuum simulation.

The set of equations to be solved is

$$\dot{\rho} = -\text{div } m , \quad \dot{m}_i = -\text{div } \phi_i \quad (i = 1, 2),$$

where

$$m = \rho V, \quad (\phi_i)_\ell = \rho u_i u_\ell + p \, \delta_{i\ell} - \mu(\frac{\partial u_i}{\partial x_\ell} + \frac{\partial u_\ell}{\partial x_i}), \quad p = \rho c^2.$$

The region of the flow is rectangular $0<x<X$, $0<y<Y$. Boundary conditions are:

at inlet ($x = 0$), $\quad m_1 = 3\bar{m}(Y^2 - y^2)/2Y^2, \quad m_2 = 0, \quad \rho = \rho_0,$

at outlet($x = X$), $\quad m_1 = 0, \quad m_2 = 0,$

at walls ($y = 0, Y$), $\quad \partial m_1 / \partial x_1 = 0, \quad \partial m_2 / \partial x_1 = 0.$

The region is covered by $8n \times 4n$ grid points, which are equally spaced on a regular mesh. The discretized equations can be written by the following conservational form.

$$\dot{f}V = [\phi_1(i-\tfrac{1}{2}, j) - \phi_1(i +\tfrac{1}{2}, j)]s_1 + [\phi_2(i , j -\tfrac{1}{2}) - \phi_2(i , j +\tfrac{1}{2})]s_2,$$

where quantities at a half-integer point are computed by

$$a_{i+\frac{1}{2}} = (a_i + a_{i+1})/2, \quad (\partial a/\partial x)_{i+\frac{1}{2}} = (a_{i+1} - a_i)/\Delta x.$$

The time march algorithm for $\dot{y} = f (y)$ by the RRK method is:

$$y^{n+1} = y^n + [g_1 g_1 / g_3]$$

where $g_1 = f(y^n)\Delta t$, $g_2 = f (y^n + cg_1)\Delta t$, $g_3 = (1-b)g_1 + bg_2,$

and $[g_1 g_1 / g_3] \equiv (2g_1(g_1 \cdot g_3) - g_3(g_1 \cdot g_1))/(g_3 \cdot g_3)$.

The parameters b and c can be chosen so as to hold bc<0.5.

Each PU calculates values of ρ, m_1, m_2 in its own region where n X n grid points are allocated. Due to the nearest neighbor property, it requires additional data on the adjacent layer of one mesh width, which are located in the 8 surrounding PUs. There is a built-in procedure that takes care of such data transfer. The transfer is made in two steps of x- and y- direction, and diagonal transfer is never necessary.

Time march by RRK method requires summation of $(\dot{\rho}c)^2 + \dot{m}_1^2 + \dot{m}_2^2$ all over the grids. This is easily made by two-step procedure: first partial summation within each PU, and then global summation of the partial sums over PUs. The latter is done by routing the partial sums in the array. The user has only to call a built-in procedure.

Performance was evaluated. Inner PUs are busier than boundary PUs. The program was analyzed to find that the number of operations such as +, -, *, /, =, and indexing is estimated by

$$C(total) = 439 + (562/n) + (619/n),$$

for one time step per grid point for the busiest PU. This value is shown in Table 3 for various values of n, i.e. problem size per a PU. The value C(total) is greater than the corresponding value of the net computation carried out in each PU,

$$C(net) = 427.$$

Most of the difference is attributable to the inter-PU data transfer, which is made 22 times in one time step.

$$C(transfer) = (528/n) + (316/n).$$

Total efficiency, C(net)/C(total), is also shown in the Table 3. This is a conservative estimate, since the data transfer is made actually faster than floating point operations, but is assumed to need equal time in this estimate.

The Table shows that the efficiency improves as the problem size increases. It is because the overhead for data transfer decreases as the "surface to volume ratio" decreases in the large problem. Processor efficiency is defined as 1 - (idle ratio of the processors), and is evaluated by analysing the work load of inner and boundary PUs, as shown in the Table 3.

Computing time for an integration step and required memory size for a PU are shown in Table 4 as a function of problem size n.

Table 3. Efficiency of parallel processing of fluid dynamics.

Problem	Operations/(Grid X Step)				Efficiency	
Size n	Net	Transfer	Other	Total	Processor	Total
4	427	152	39	618	0.941	0.691
8	427	71	21	519	0.962	0.823
16	427	38	15	480	0.975	0.890
32	427	17	13	457	0.988	0.934

Table 4. Time and memory requirement

Problem Size n	Computing time(s)/Step	Memory(KB)/PU
4	0.8	3.7
8	2.5	10.8
16	8.5	35
32	31.9	126

Here the scaling law .is derived. The efficiency turns out to range in 70-90%. It is dependent on n, number of grid points per PU, and is independent of number of processors. Thus we can conclude that PAX will have a powerful performance proportional to the number of processors, unless it exceeds the required grid size. Such a case would never happen, because hydrodymamicists ordinarily face with the problems of grid size more than 10000 X 10000 X 10000.

6. Future Super Freedom Simulator PAX-128, PAX-4096 and Beyond.

The linear scaling law ensures that the very large scale VLSI array of PUs still works with the same efficiency as that evaluated on PAX-32. However, before jumping to the goal, a super freedom simulator PAX-4096, we must take several intermediate steps to make sure the feasibility and to improve the hardware design.

Our next step, now being constructed, is PAX-128. The machine consists of 128 PUs, that follows almost the same design as PAX-32, but includes some improvements and simplifications in the memory structure. The PU array is implemented in a torus. The torus shape makes the interprocessor connection minimum. The performance of PAX-128 is predicted as 4 MFLOPS, that is equal to 128 times 0.03 MFLOPS. The cost is estimated as about \$30 000.

The goal PAX-4096 consists of 4096 VLSI PUs. We can expect 1 MFLOPS for the VLSI performance, so that PAX-4096 will have 4 GFLOPS speed. The memory size of a few 100 Kbytes is desirable for a single PU. The array is a 2-dimensional 64 X 64 square with the end-around connection.

The device technology to be used in PAX-4096 is not clear at the present stage. The performance of the system is measured in terms of the unit speed times number of units, so that the high density in integration and implementation is as important as the unit speed. The recent remarkable progress of CMOS technology is very promising in terms of this performance·criterion. The cost for the initial investment on the VLSI is estimated as a few million dollars. The ratio of performance over the cost is about 10 to 100 times better than the conventional computers. The reliability is another limiting factor of this type of architecture. Based on the recent figure, 1000 fit/chip, the system that consists of 4096 PUs X 3 chips/PU has the MTBF of 80 hours. That may be acceptable in comparison with the computational speed of 4 GFLOPS. As for the heat dissipation, it will be some 10 Kwatt if we use CMOS devices.

The extension to the further step: probably PAX-1M system, with 1 million PUs and 1 TFLOPS(Tera FLOPS) of total performance, will be disccused in 1990s as the device technology progresses.

7. Conclusions

Super freedom simulator makes the physics work, not only in the fundamental aspects but also in the practical engineering problems, by simulating the physical process by the numerical means. The simulator has a power enough to cope with the immense needs of computational speed. Actually the computational power of 1 GFLOPS or more is a crucial factor in the development of new energy resources such as nuclear fusion.

The pilot machine PAX-32 demonstrated that the nearest neighbor mesh processor array of MIMD type is useful in the typical examples of the scientific calculations. The examples cover not only the partial differential equation models, but also the random spin system with large degrees of freedom. The computer system versatile in both application models can be said of general use in the field of scientific calculation. It seems there is no need to develop more complex architecture aiming at more generally usable machine, with the cost of hardware and software difficulties.

The architecture of PAX is such that large number but a few types of VLSIs are regularly implemented in the array.   This fits in most with the VLSI technology. The simulator will have a big market throughout the world, if it is tailored as a special-purpose simulator with built-in program package.   The simulator will have a better cost performance than the conventional "super computer," that is for millionairs.   In this sense, the simulator may be compared to the mini- or micro-computers.

The VLSI technology put the new trend in the computer design.   The technology is still the most important factor for the feasibility of the large scale simulator; especially the ease of development of the custom LSI is crucial.

References
[1]. Hoshino T., Kawai T., Shirakawa T., Higashino J., Yamaoka A., Ito H., Sato T., and Sawada K., PACS, A Parallel Microprocessor Array for Scientific Calculations, submitted to Transactions on Computing Systems, ACM (1982).
[2]. Wilson, K., Phys. Rev. D10, 2445 (1974).
[3]. Satofuka, N., A New Explicit Method for the Numerical Solution of Parabolic Differential Equations, Proc. 2nd National Symp. on Numerical Methods in Heat Transfer, Univ. of Maryland (Sept. 1981).
[4]. Wambecq A., Rational Runge-Kutta Method for Solving Systems of Ordinary Differential Equations, Computing 20, 333-342 (1978).

PROLOG INTERPRETER AND ITS PARALLEL EXTENSION

Yuji MATSUMOTO   Katsumi NITTA   Koichi FURUKAWA

Electrotechnical Laboratory   Institute for New Generation
                                     Computer Technology
Ibaraki, Japan, 305   Tokyo, Japan, 108

1. Introduction

Prolog is a computer language based on predicate logic. Kowalski [1] first showed that procedural interpretation of Horn clauses, which are a subset of first order predicate logic, could be used as an efficient programming language. Prolog was first implemented in Marseille [2]. The first compiler was implemented in Edinburgh together with an interpreter and it has been shown that its efficiency is comparable to LISP [3].

This paper presents a Prolog interpreter based on concurrent programming and its extension to a parallel execution model. The execution of a Prolog program can be seen as a depth first traverse of an AND/OR tree defined by the program. Our interpreter is based on this point of view, and consists of two kinds of processes, AND-process and OR-process.

Since Prolog is a language based on predicate logic, the order to interpret a program is intrinsically implementation oriented. Usually the strategy adopted by the Prolog implementation is SNL (selective negative linear). Our interpreter also obeys this strategy. There has been considered possibilities of some types of parallel interpretation of Prolog programs, and Prolog in these parallelism naturally demands the programmer the different form of Prolog programs, that is, a program must be independent of the sequence of the clauses or the sequence of the literals in the clause.

Our parallel execution model, called backup parallelism, is a natural extension of our AND/OR process model of Prolog interpreter. Each process corresponds to a processor. The backup parallelism means that there is one main process and other processes work for the job needed when the backtracking occurs. There is no different operation between the main process and the others. Processors are dynamically allocated to the processes.

Next chapter outlines our Prolog interpreter based on AND/OR process model. The behavior of the interpreter for a sample program is shown and two major ways to deal with data structures in Prolog is discussed, of which we adopted the Copy method.

Chapter 3 and 4 present the backup parallelism which is the natural extension of our Prolog interpreter. This parallel model executes the same Prolog program as the sequential one and requires less execution time.

Chapter 5 shows the simulation and experiments of our backup parallel model. The simulator written in SIMULA is introduced and some examples are given.

2. Sequential Prolog Interpreter by Concurrent Programming

2.1. Control Part

A Prolog program is a set of statements called clauses. The left side and the right side of a clause are called "head" and "body", respectively. A body consists of a sequence of more than zero goals. The clause without head is an initial clause. An initial clause triggers the interpretation.

The execution of Prolog program corresponds to the search on an AND/OR tree. For example, suppose the following program is given.

```
1. a(X) <- b(X),c(X).
2. b(X) <- e(X),f(X).
3. b(X) <- g(X).
4. c(X) <- h(X).
5. e(dick) <-.
6. f(dick) <-.
7. g(tom) <-.
8. h(tom) <-.

0. <-a(X).   ! initial clause ;
```

When this program is executed, the interpreter works as if it traverses the AND/OR tree given in Fig.1, in left-to-right and depth-first manner. On search, variables in a clause are instantiated, and the result is used as the environment at the new instantiation process called "unification".

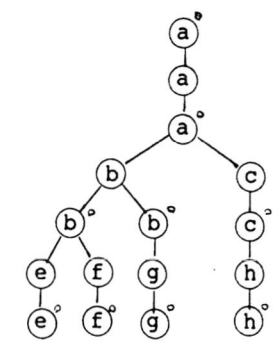

○ : and-node ◯ : or-node

FIG.1 AND-OR TREE

Our interpreter consists of two parts called AND-process and OR-process. An AND-process behaves the same function as the AND node on the AND/OR tree. An OR-process behaves the same function as the OR node.

When initial clause is given and interpreter starts, an AND-process is created and activated. The AND-process gets a sequence of goals in the initial clause as an argument. For the leftmost goal of the sequence· the AND-process creates an OR-process, passes the control to it and waits for the result. When the son OR-process returns the result· the AND-process is activated. If the result is SUCCESS, the OR-process is pushed into B-stack of the AND-process, and then the AND-process creates an OR-process for the next goal and repeats the same behavior. When the OR-process of the last goal succeeds, the AND-process returns the result SUCCESS to the main part of interpreter. If the son OR-process fails, the AND-process pops up an OR-process from B-stack, and reactivate it for another solution. If it succeeds, the interpretation is executed from the next goal of the OR-process. If it fails, an OR-process is popped and reactivated again. When the B-stack is empty and the AND-process cannot pop up any OR-process, the AND-process returns FAILURE.

An OR-process is created by an AND-process and is given a goal as an argument. The OR-process searches a clause Ck whose head is the same as the goal, and executes the unification. If unification succeeds, the OR-process creates an AND-process for the body of Ck, passes the control to it and waits for the result. If the son AND-process returns SUCCESS, the OR-process is activated and returns SUCCESS to the parent AND-process. When the unification fails or the son AND-process fails, the OR-process searches another clause and repeats the same behavior. If the OR-process tries all candidate clauses and does not succeed, the OR-process fails.

The control transfers from a parent process to a son process or from a son to a parent, and only one process which is called as main process is active at a time, therefore each process behaves as a coroutine. When AND-process and OR-process return SUCCESS, they are suspended and preserve the state, preparing backtracking. When they fail, they are deleted.

When the example program is executed, the relation between created processes is shown in Fig.2. In figure (a), OR-process O6 fails and the backtrack occurs and processes O6, A2, O3, A3, O4, A4, O5 and A5 are deleted.

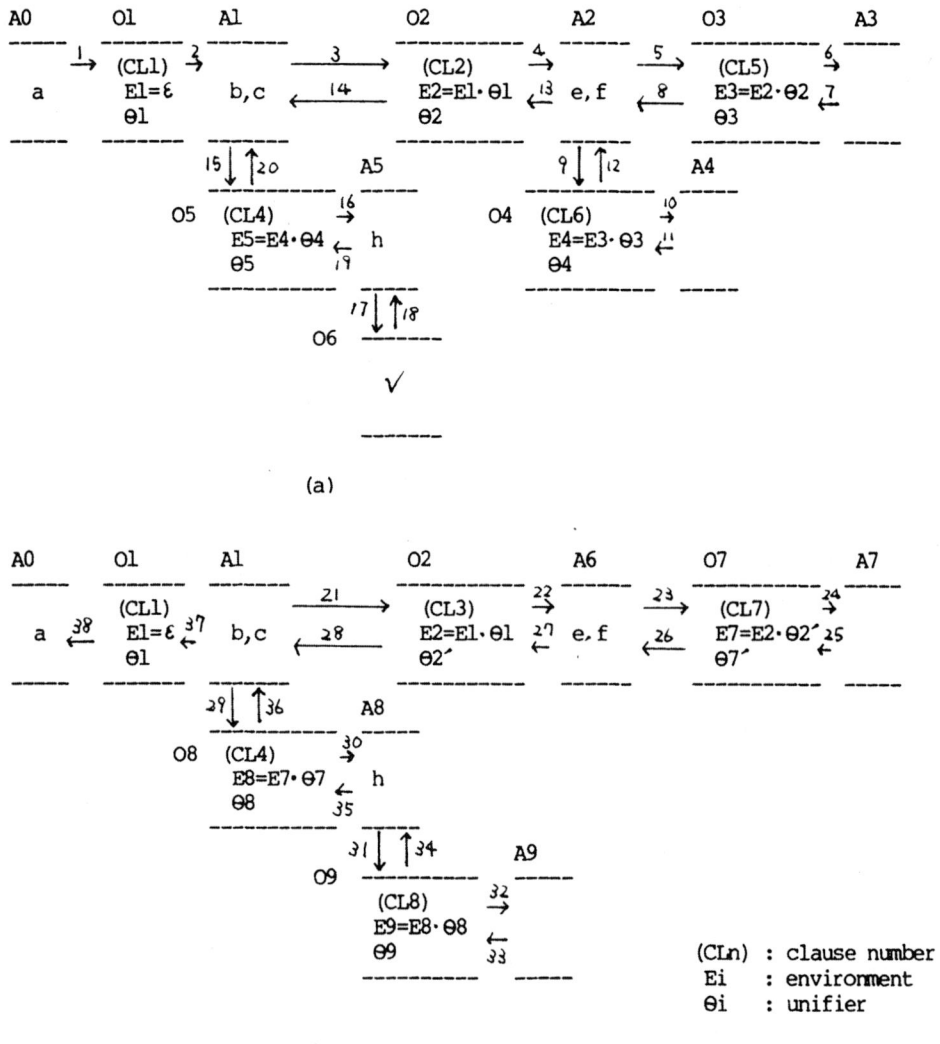

(a)

(b)

(CLn) : clause number
Ei   : environment
θi   : unifier

FIG.2 GENERATED PROCESSES

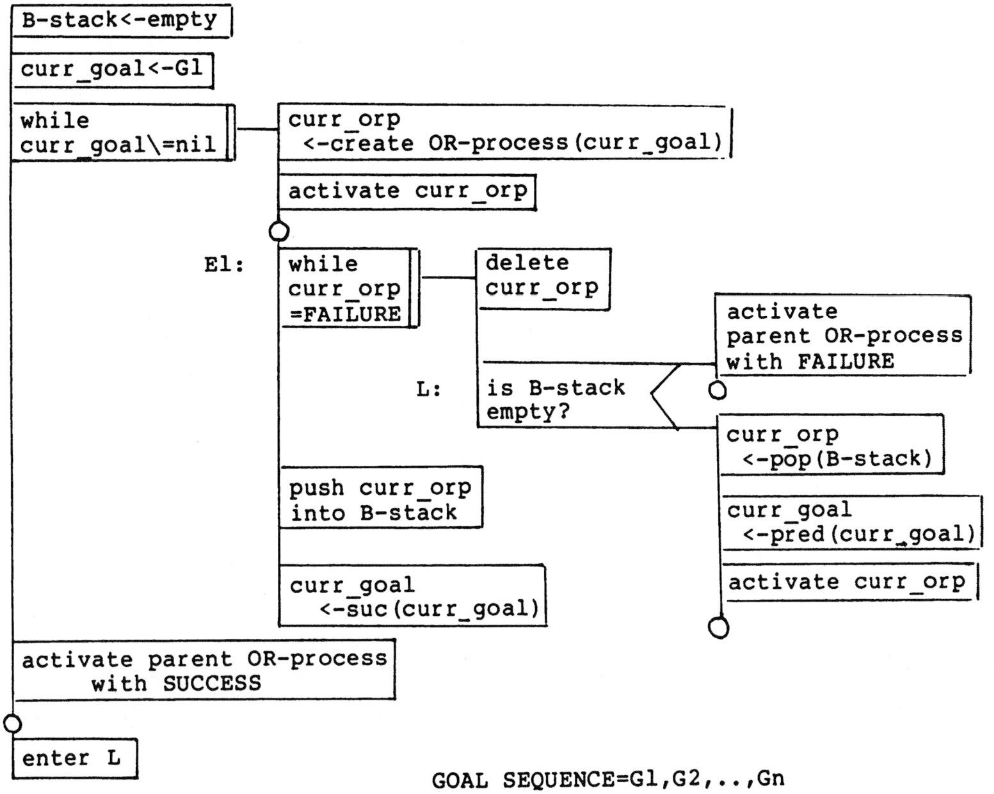

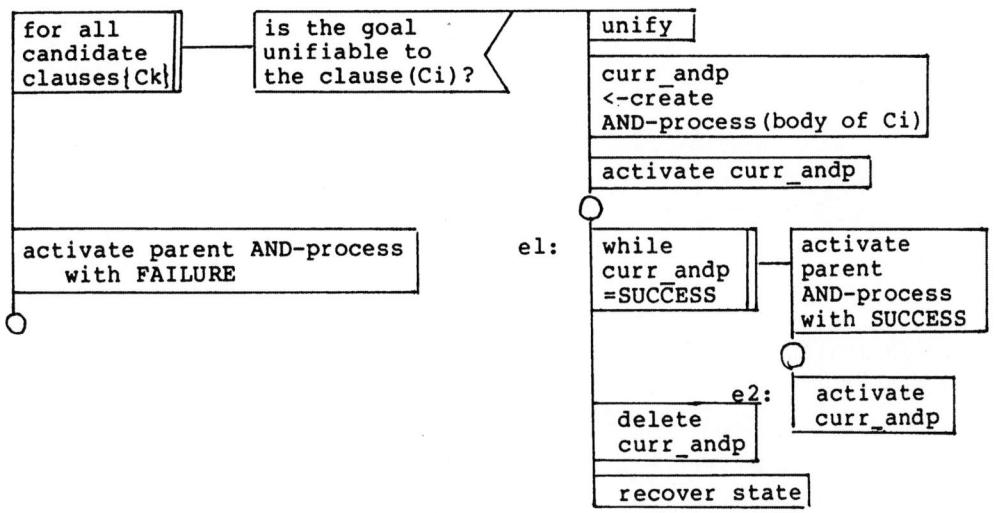

```
B-stack<-empty

curr_goal<-G1

while              curr_orp
curr_goal\=nil      <-create OR-process(curr_goal)

                   activate curr_orp

        E1:  while        delete
             curr_orp     curr_orp
             =FAILURE                      activate
                                           parent OR-process
                                           with FAILURE
                   L:  is B-stack
                       empty?             curr_orp
                                           <-pop(B-stack)
             push curr_orp
             into B-stack                 curr_goal
                                           <-pred(curr_goal)

             curr_goal                    activate curr_orp
              <-suc(curr_goal)

activate parent OR-process
     with SUCCESS

enter L
```

GOAL SEQUENCE=G1,G2,..,Gn

FIG.3   AND-PROCESS(GOAL SEQUENCE)

```
for all         is the goal          unify
candidate       unifiable to
clauses{Ck}     the clause(Ci)?      curr_andp
                                      <-create
                                      AND-process(body of Ci)

                                     activate curr_andp

activate parent AND-process    el:   while         activate
     with FAILURE                    curr_andp     parent
                                     =SUCCESS      AND-process
                                                   with SUCCESS

                                              e2:   activate
                                     delete          curr_andp
                                     curr_andp

                                     recover state
```

FIG.4   OR-PROCESS(GOAL)

The flow charts of AND-process and OR-process are shown in Fig.3 and Fig.4. They are depicted by PAD (Problem Analysis Diagram) [4]. Circles in the figures mean waiting states. (In the figures, only the fundamental part is shown, and optimizing function and cut operation are omitted.)

2.2. Unification Part

An OR-process has slots which contain instances of variables in the clause selected by the OR-process. An AND-process can access the slots in the parent OR-process. When an unification is executed in an OR-process, the result is registered in slots in the OR-process and the parent AND-process.

Unification is usually executed on Robinson's algorithm [5], and there are two methods for data representation, called Structure Sharing and Copy. In Structure Sharing method, instances of variables are expressed by a couple of pointers << pointer to literal, pointer to environment to evaluate the literal >>. On getting the value of a variable, a chain of pointers must be traced and the required structure is to be constructed. In Copy method, when the instance literal has variables within it, its copy is constructed in copy stack and a pointer to it is put into the slot. When the instance literal has no variable, the slot directly points the literal.

Examples of both methods are shown in Fig.5. This figure shows the snapshot of the execution of the following APPEND program.

```
1. append(.(W,X),Y,.(W,Z)) <- append(X,Y,Z).
2. append(nil,X,X) <-.

0. <- append(.(a,.(b,nil)),.(c,.(d,nil)),P).
```

In a sequential interpreter, both methods have advantages and there is little difference in the amount of required memory [6],[7]. However, in a parallel interpreter explained in the following sections, memory conflict must be considered carefully. We select Copy method in our interpreter, because the method is easy to apply to our process-based interpreter and Structure Sharing is weak in the memory conflict.

58

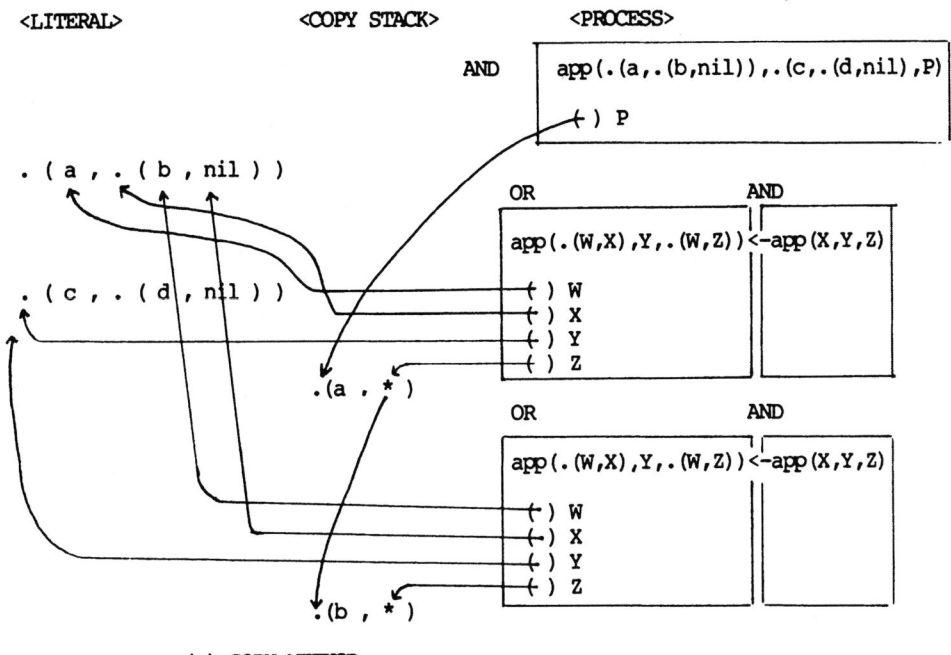

(a) COPY METHOD

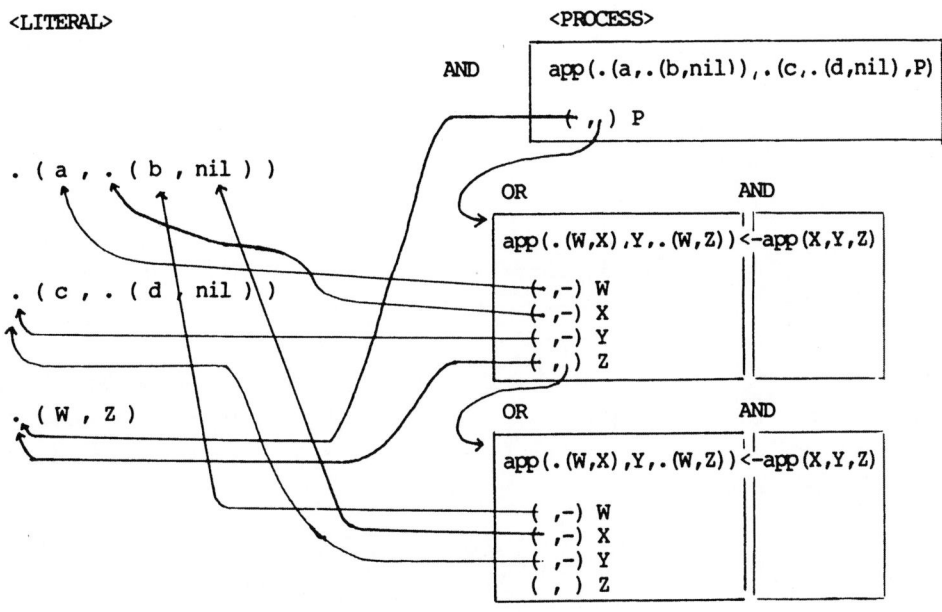

(b) STRUCTURE SHARING

FIG.5 SNAPSHOT OF APPEND PROGRAM

3. Parallel Realization of Prolog Interpreter

3.1. Parallelisms in Prolog

Since Prolog is a language based on predicate logic, the order to solve goals is not restricted intrinsically. There can be considered two basic types of parallel control flow in a Prolog program. They are AND-parallelism and OR-parallelism. Suppose we have the clause h(X,Y):-p(X,Y),q(X),r(Y). When Prolog tries to solve a goal with the head h and this clause unifies with it, the body of this clause becomes the new goals. This clause is solved if and only if each of the goals in its body is solved.

Three goals in the body are connected by logical conjunction. From the viewpoint of declarative semantics, they can be solved in any order. AND-parallelism insists on the simultaneous execution of these goals. There is a difficulty inherent in AND-parallelism. A variable included in more than one goal in a clause body must have the same value throughout the computation. For example, p(X,Y) and q(X) in the body have the same variable X. If these are executed in the same time and one of them gets a value assignment to the variable X, this limits the value of the variable in the other goals to the same value. More precisely, each of the goals eventually gets the set of all combinations of the variable assignment and the whole answer of the clause is to be the intersection of the sets for the goals in the body.

When p(X,Y) is called, there may be more than one clause whose head unifies with · this call. OR-parallelism insists on the simultaneous execution of all the unifiable clause bodies. Since variables in different clauses are mutually independent, there is no difficulties in OR-parallelism as in AND-parallelism. However, assigning one processor to each OR-process leads the explosion of the number of the processors.

3.2. Backup Parallelism in Prolog

This section shows a form of parallelism as we call "backup parallel execution" of Prolog programs.

In the preceding section, we have indicated some types of parallelisms in Prolog. These parallel execution neglect the order of the goals in a body or the order of the clauses in a program. When a given program has several answers, the programmer cannot predict the order of the answers obtained from these parallel execution. In case there are cut symbols in the program, these order should not be neglected. Moreover, users of Prolog usually write their program in sequential order, and writing sequential program is easier for them.

The backup parallel execution considered here is the natural extension of our AND/OR model of Prolog interpreter.
In our AND/OR model a parent AND(OR)-process creates and activate a OR(AND)-process, and the son process answers the solution when it finds one. In the backup parallel execution, the son process tries to

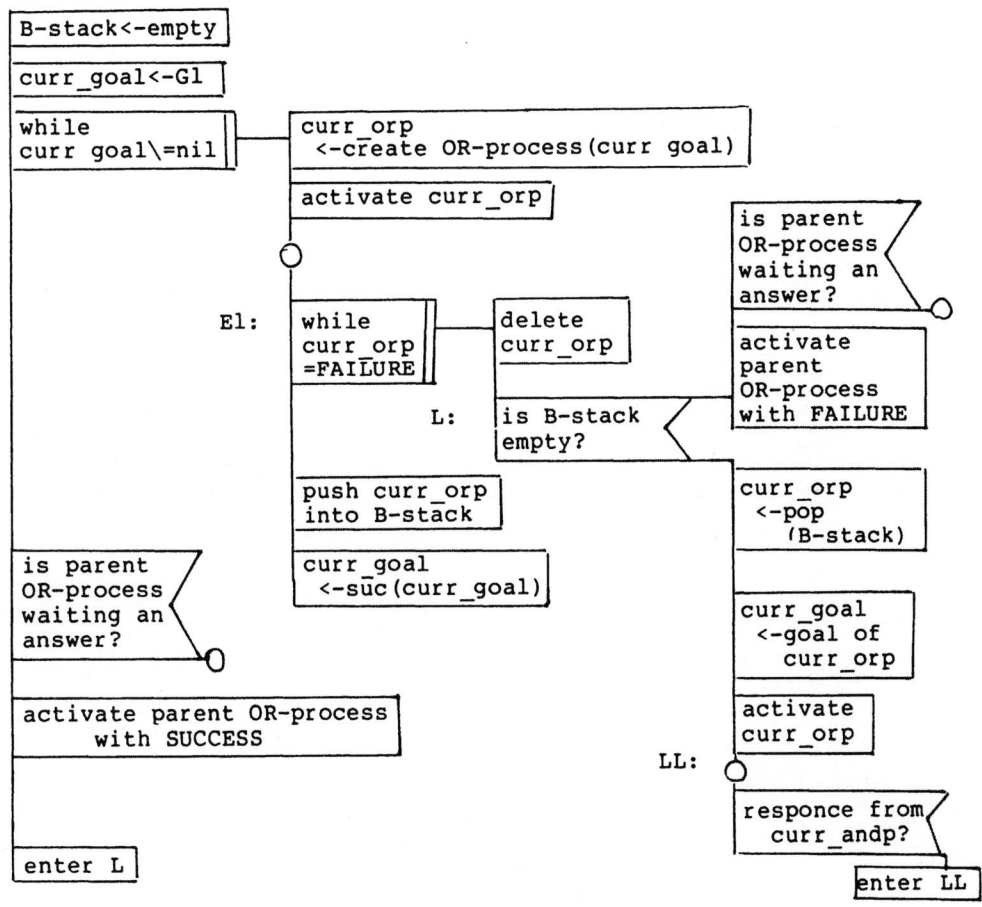

GOAL SEQUENCE=G1,G2,..,Gn

FIG.6   AND-PROCESS(GOAL SEQUENCE)

find another answer as soon as it sends an answer to the parent process. A process is suspended when it finds another answer and the parent process is not waiting for the answer. A process in this state can answer the solution immediately after the reactivation from the parent process. A process which has failed to find the answer send a failure message to the parent and can be deleted. The operations of AND-process and OR-process in the backup parallel execution are depicted in Fig.6 and Fig.7, respectively.

An AND-process deals with a sequence of goals. It creates an OR-process for the first goal and activates it. After the activation, it waits for an answer from the son OR-process. If the son sends back a solution, the next goal is picked up under this environment to create the next OR-process. If the answer from the son OR-process is failure, it reactivates the previous OR-process.

If the OR-process dealing with the last goal sends a success answer, it sends the success answer, which consists of the whole

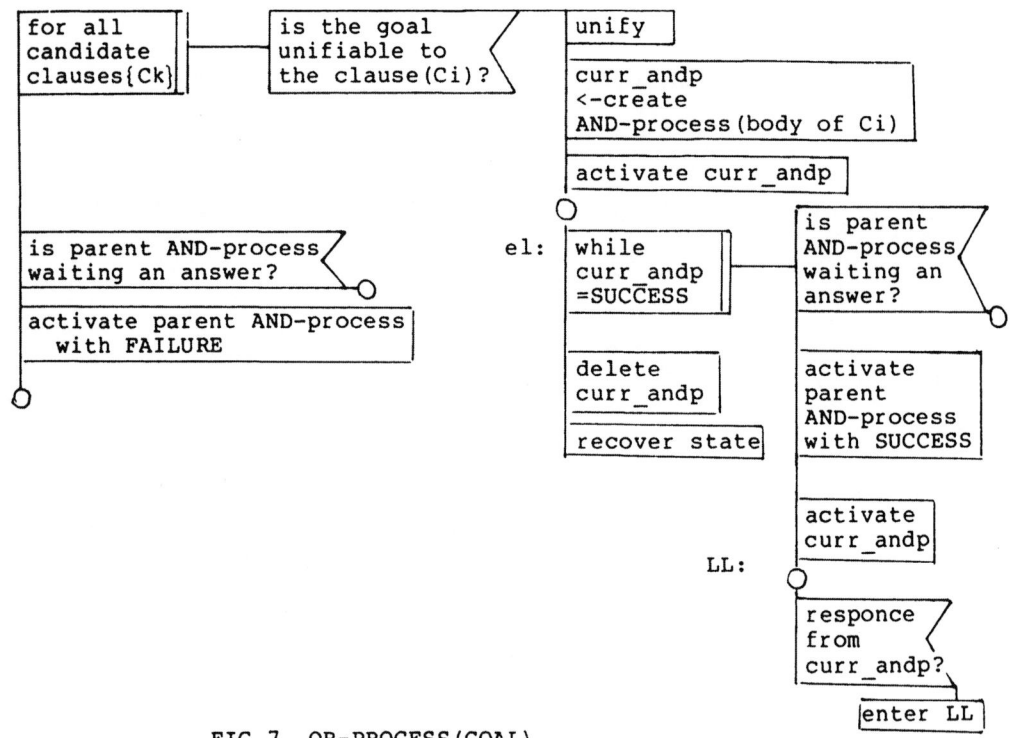

FIG.7   OR-PROCESS(GOAL)

variable bindings from the son OR-processes, to the parent OR-process. Immediately after it sends the answer to the parent OR-process, it reactivates the last son OR-process to obtain another solution. If this process gets another solution before the reactivation from the parent process, it enters the full mode.

An AND-process fails when the first son OR-process fails to find the answer.

An OR-process corresponds to a goal in the parent AND-process and is created and activated by that AND-process. It first searches a clause whose head unifies with itself, creates the AND-process corresponding to the body and activates it. If the son process answers with success · this solution is sent to the parent AND-process. If the answer from the son process is failure, another clause whose head unifies with the original goal is searched and the body of that clause is activated as the new son AND-process.

Immediately after sending the solution to the parent AND-process, it reactivates the son AND-process to obtain another solution. When it gets another solution before the reactivation from the parent process, it enters the full mode.

The OR-process fails when it gets failure from the son process and there is no other clause whose head unifies with the original goal.

Although we have assumed in the preceding explanation that a process corresponds to a processor and each process searches only one extra solution, there are some alternatives.

If each process has an infinite buffer to keep solutions, the process can activate the son process repeatedly until it gets all answers. Our previous model is a case where the length of the buffer is 1. The reason to limit the length to 1 is for the simplicity of the simulator. If the length of the buffer is more than one, each process enters the full mode when the buffer becomes full.

We have distinguished AND and OR process in the preceding discussion. However, an OR-process deals with the head of a clause at a time and its son process works with the body of the clause. Since variables are shared within a clause, it is reasonable to allocate a processor to a pair of OR-process and its son AND-process. Our simulator for the backup parallel model, shown in Chapter 5, has been constructed under this consideration.

Although an AND-process usually has several sons, it communicates with only one son process, which is on the top of the B-stack in the AND-process. Suppose that a son OR-process whose corresponding goal is in the B-stack of its parent AND-process fails to find a solution. In the model stated above, the OR-process keeps the failure answer and wait for the parent's reactivation. However, if the preceding OR-process has already found a solution and this information can be sent to the OR-process, this OR-process can start to find solutions in this new environment. Such information transferring between son OR-process leads more efficient backup parallelism. Our simulator has not yet been facilitated with this ability.

4. Data Structure for Backup Parallelism

As shown in section 2.2, there are two major data representation methods in Prolog implementation. Of these we adopted Copy method for our Prolog interpreter. When unification occurs, a structure which includes undefined variables is copied.

In the backup parallel execution, each process works in its own environment. A variable in a Prolog program may have different value assignments in different processes. Therefore, a variable must be represented by a set of values, such that each process can access the proper value of them.

We represent a variable by the pair << generation , value >> in the backup parallel model, where the generation shows when this value is assigned to the variable. Hence a variable consists of a list of such pairs. In our parallel interpreter, each generation is represented by a number which indicates a unification. A process has a set of numbers which is the history of the unifications this process concerns. Our parallel simulator directly uses this way of variable representation. We have mentioned in Section 2.2 that we adopt Copy method for data representation. When this method is used in our parallel model, structures including undefined variables are copied. In the case of sequential execution, messages passed between parent and son processes are only SUCCESS or FAILURE, for the solution can be obtained through the bindings to the variables made by the unification. In the parallel case, however, the set of generation is sent to the parent to construct the solution. Although copied structures are not shared by the variables in the same generation, they may be shared by the variables in the different generations.

Unfortunately, we have not yet decided an efficient realization of the above data representation. This is the key of the efficiency of our backup parallel model. At least the main process should be able to access its environment as fast as the process in the sequential model.

5. Simulation and Experiments

We implemented the Prolog interpreter shown in previous section on concurrent language SIMULA. To measure the execution time, in each step of AND/OR-process, we give the time to be required as parameters.
The interpreter has two versions.
(1) the sequential interpreter (SI) which is the implementation of Fig.3 and 4.
(2) the parallel interpreter (PI) which is the implementation of Fig.6 and 7.
If infinite processors are available, every pair of OR-process and its son AND-process is allocated to one processor and can work its job communicating with its parent or its son process. However, in case the number of processors is bounded, the number of processes often exceeds that of processors and some process must wait until a processor is released.
Our interpreter PI does not need so many processors comparing with processors required by AND-parallel or OR-parallel interpreter. However, when the execution of a program backtracks many times, PI consumes many processors.

The allocation of processor is done as following.
Each process corresponds to a node on AND/OR tree, and a measure of priority is attached to the node. The priority is settled when the AND/OR tree is traversed in left-to-right and depth-first manner. The farther the node is from the root, the higher priority is attached to the node. In our interpreter PI, a main process has a initiative and other processes work to back it up, which is a different point from AND-parallel or OR-parallel. Therefore, when processors are not enough, the process whose priority is lowest among active processes is enforced to be into the waiting state and pushed in the waiting process stack. When some processor is released, a process whose priority is the highest among processes in the stack is popped and allocated with this processor.

As the example program, we use the following permutation program.

```
1. perm(nil,nil) <-.
2. perm(.(X,Y),.(U,V)) <- del(U,.(X,Y),W),perm(W,V).
3. del(X,.(X,Y),Y) <-.
4. del(U,.(X,Y),.(X,V)) <- del(U,Y,V).

0. <- perm(.(1,.(2,.(3,nil))),RESULT).
```

This program has six solutions such as (1 2 3), (1 3 2), .., (3 2 1). The execution time of SI and PI are shown in Table 1. (The case where the number of processors is bounded is not shown here.)

	SI	PI
(1 2 3)	245	244
(1 3 2)	542	379
(2 1 3)	986	590
(2 3 1)	1283	725
(3 1 2)	1741	936
(3 2 1)	2038	1071
finish	2340	1110

Table 1.

The time unit of this table is system time of SIMULA and corresponds to the real execution time. The time required to find first solution (1 2 3) is almost the same, for backtrack does not occur. However, the time to find other solutions is decreased. In the execution of PI, the number of allocated processors changes dynamically. The rates of time in which one, two, three and four processors are allocated are about 40%,40%,15% and 5%, respectively.

6. Concluding Remarks

A Prolog interpreter and its extension to Parallel execution model are introduced. Our parallel model is called backup parallel, which works on the AND/OR tree constructed by a Prolog program. This parallel model can run the original sequential Prolog program faster with processors of the same type.

As shown in Section 5, this model can earn a few times faster execution time with several processors, provided that the order of solutions given by the parallel interpreter is the same as that of sequential one. Programmers does not need to worry about the control of the interpreter.

There are some problems to be solved. One is the processor allocation and the other is the data representation.

When the number of the processors is limited, processors should be allocated to the nearest processes to the main process. Difficulty is not located on which are the nearest processes, because processes can be ordered according to the history they are created. In the parallel execution, processes are created dynamically not only by the main process. However, the order of the son process is due to the order of the parent process and it can be reordered into other processes, which creates the total order of the processes. When a process enters into the waiting mode or the full mode, the processor is released from the process. A processor is allocated to such a process when it is activated by son or parent process and there are free processors or there is a process not nearer to the main process.

We adopted Copy method for our interpreter. It is also utilized in our parallel model. A variable in a Prolog program is represented by a pair of its generation and the value. A variable is accessed by several processes which work under their own environments. They access the value of the variable by the generation it belongs to. It is the remained and most important problem how to realize the data representation which can be accessed efficiently by several processors.

References

[1] Kowalski, R., "Predicate Logic as Programming Language,"
    Information Processing 74, North-Holland, pp.569-574, 1974.
[2] Roussel, P., "Prolog: Manual de Reference et d'Utilisation,"
    Groupe d'Intelligence Artificielle, Marseille-Luminy, Sept. 1975.
[3] Warren D.H.D. et al., "Prolog-The Language and its Implementation
    Compared with Lisp," Proc. Symposium on Artificial Intelligence
    and Programming Languages, pp.109-115, 1977.
[4] Futamura, Y. et al., "Development of computer programs by PAD,"
    Proc. of the Fifth International Conference on Software
    Engineering, 1982.
[5] Chang, C. L. and Lee, R. C., "Symbolic Logic and Mechanical Theorem
    Proving," Academic Press, 1974.
[6] Bruynooghe, M., "The Memory Management of PROLOG Implementation,"
    Logic Programming Workshop, 1980.
[7] Mellish, C.S., "An Alternative to Structure-Sharing in the
    Implementation of A Prolog Interpleter," Logic Programming
    Workshop, 1980.

Chapter 3

VLSI Algorithms

Putting Inner Loops Automatically in Silicon

H. T. Kung

Department of Computer
Carnegie-Mellon University
Pittsburgh, Pennsylvania 15213
U.S.A.

Abstract

Many of the time consuming inner loops are inherently regular and parallel. These are exactly the structures that are well suited for VLSI implementation. As a result, it will become increasingly common to have subroutines that are directly executeable in silicon. Does it imply that in the near future many large computations can be effectively carried out by small computers equipped with silicon subroutines? This talk will present a simplied characterization of the silicon subroutine approach, and discuss systolic architectures—a powerful method for implementing cost-effective silicon subroutines for computations such as pattern matching and error-correcting. CAD systems at CMU that have made it possible for us to design some rather complex chips, such as a programmable systolic chip, will also be briefly described.

The research was supported in part by the Office of Naval Research under Contracts N00014-76-C-0370, NR 044-422 and N00014-80-C-0236, NR 048-659, and in part by the Defense Advanced Research Projects Agency (DOD), ARPA Order No. 3597, monitored by the Air Force Avionics Laboratory under Contract F33615-81-K-1539.

1. Introduction

An emerging belief among many researchers is that a significant portion of the next generation of high performance computers will be based on architectures capable of exploiting very large scale integration (VLSI) modules. In particular, it is desirable to have a compact and inexpensive "hosts" that can be plugged in with interchangeable high performance modules to fit various application requirements. Such a host can be an efficient signal processor when special-purpose signal processing modules are used; it can also be an efficient database machine when the modules are replaced with data processing modules.

CMU has in recent years been actively involved in research concerning special-purpose VLSI designs. Many algorithms and architectures that are well suited for VLSI implementation have been proposed, and prototypes have been implemented in chips. More importantly, a design methodology, *systolic architectures*, has been evolved. Using the systolic approach, many important functional modules in application areas such as signal and image processing, as well as in database processing, can be mapped into silicon cost-effectively. From these results and other advances, such as fast FFT chips that are becoming commercially available, we feel that design and implementation costs of many special-purpose devices are no longer a major difficulty. Section 2 reviews the basic concept of systolic architectures and carefully explains why they should result in cost-effective, high-performance special-purpose systems applicable to a wide range of problems. A family of systolic arrays for the convolution problem are illustrated in Section 3. Section 4 is a project overview of a programmable systolic chip that is under development at CMU. In Section 5 we discuss system issues such as convenient incorporation of these devices into complete systems, and their effective utilization from a system point of view.

2. Fundamental Principle of Systolic Architectures

In a systolic system, data flows from the computer memory in a rhythmic fashion, passing through many processing elements before it returns to memory, much as blood circulates to and from the heart. In some sense, the system works like an automobile assembly line where different people work on the same car at different times and many cars are assembled simultaneously. An assembly line is always linear, however, and systolic systems are sometimes two-dimensional. They can be for example rectangular, triangular, or hexagonal to make use of higher degrees of parallelism. Moreover, to implement a variety of computations, data flow in a systolic system may be at multiple speeds in multiple directions—both inputs and (partial) results flow, whereas only results flow in classical pipelined systems. Since each data is used multiple times, systolic architectures can speed up execution of compute-bound problems without increasing I/O requirements. An additonal advantage of a systolic system is that it is easy to implement because of its regularity and to reconfigure (to meet various outside constraints) because of its modularity.

The systolic architectural concept was developed at Carnegie-Mellon University [6, 13, 20, 21, 25, 26, 28, 30], and versions of systolic processors are being designed and built by several industrial and governmental organizations [4, 7, 45, 49].

2.1 Key Architectural Issues in Designing Special-Purpose Systems

Roughly, the cycle for developing a special-purpose system can be divided into three phases — task definition, design, and implementation. During task definition, some system performance bottleneck is identified, and a decision on whether or not to resolve it with special-purpose hardware is made. The evaluation required for task definition is most fundamental, but since it is often application dependent, we will concentrate only on architectural issues related to the design phase, and will assume routine implementation.

2.1.1 Simple and Regular Design

Cost-effectiveness has always been a chief concern in designing special-purpose systems; their cost must be low enough to justify their limited applicability. Costs can be classified as nonrecurring (design) and recurring (part) costs. Part costs are dropping rapidly due to advances in integrated circuit technology, but this advantage applies equally to both special-purpose and general-purpose systems. Furthermore, since special-purpose systems are seldom produced in large quantities, part costs are less important than design costs. Hence, the design cost of a special-purpose system must be relatively small for it to be more attractive than a general-purpose approach.

Fortunately, special-purpose design costs can be reduced by the use of appropriate architectures. If a structure can truly be decomposed into a few types of simple substructures or building blocks, which are used repetitively with simple interfaces, great savings can be achieved. This is especially true for VLSI designs where a single chip comprises hundreds of thousands of components. To cope with that complexity, simple and regular designs, similar to some of the techniques used in constructing large software systems, are essential [34]. In addition, special-purpose systems based on simple, regular designs are likely to be modular and therefore adjustable to various performance goals — that is, system cost can be made proportional to the performance required. This suggests that meeting the architectural challenge for simple, regular designs yields cost-effective special-purpose systems.

2.1.2 Concurrency and Communication

There are essentially two ways to build a fast computer system. One is to use fast components, and the other is to use concurrency. The last decade has seen an order of magnitude decrease in the cost and size of computer components but only an incremental increase in component speed [37]. With current technology, tens of thousands of gates can be put in a single chip, but no gate is much faster than its TTL counterpart of 10 years ago. Since the technological trend clearly indicates a diminishing growth rate for component speed, any major improvement in computation speed must come from the concurrent use of many processing elements. The degree of concurrency in a special-purpose system is largely determined by the underlying algorithm. Massive parallelism can be achieved if the algorithm is designed to introduce high degrees of *pipelining* and *multiprocessing*. When a large number of processing elements work simultaneously, coordination and communication become significant — especially with VLSI technology where routing costs dominate the power, time, and area required to implement a computation [43]. The issue here is to design algorithms that support high degrees of concurrency, and in the mean time to employ only simple, regular communication and control to enable efficient implementation.

2.1.3 Balancing Computation with I/O

Since a special-purpose system typically receives data and outputs results through an attached host, I/O considerations influence overall performance. (The host in this context can mean a computer, a memory, a real-time device, etc. In practice, the special-purpose system may actually input from one "physical" host and output to another.) The ultimate performance goal of a special-purpose system is — and should be no more than — a computation rate that balances the available I/O bandwidth with the host. Since an accurate a priori estimate of available I/O bandwidth in a complex system is usually impossible, the design of a special-purpose system should be modular so that its structure can be easily adjusted to match a variety of I/O bandwidths.

Suppose that the I/O bandwidth between the host and a special-purpose system is 10 million bytes per second, a rather high bandwidth for present technology. Assuming that at least two bytes are read from or written to the host for each operation, the maximum rate will be only 5 million operations per second, no matter how fast the special-purpose system can operate. (See Figure 2-2 in the next section.) Orders of magnitude improvements on this throughput are possible only if multiple computations are performed per I/O access. However, the repetitive use of a data item requires it to be stored inside the system for a sufficient length of time. Thus, the I/O problem is related not only to the available I/O bandwidth, but also to the available memory internal to the system. The question then is how to arrange a computation together with an appropriate memory structure so that computation time is balanced with I/O time.

The I/O problem becomes especially severe when a large computation is performed on a small special-purpose system. In this case, the computation must be decomposed. Executing subcomputations one at a time may require a substantial amount of I/O to store or retrieve intermediate results. Consider, for example, performing the n-point fast-Fourier-transform (FFT) using an S-point device when n is large and S is small. Figure 2-1 depicts the n-point FFT computation and a decomposition scheme for $n = 16$ and $S = 4$. Note that each subcomputation block is sufficiently small so that it can be handled by the 4-point device. During execution, results of a block must be temporarily sent to the host and later retrieved to be combined with results of other blocks as they become available. With the decomposition scheme shown in Figure 2-1 (b), the total number of I/O operations is $O(n \log n / \log S)$. In fact, it has been shown that, to perform the n-point FFT with a device of $O(S)$ memory, at least this many I/O operations are needed for any decomposition scheme [18]. Thus, for the n-point FFT problem an S-point device cannot achieve more than an $O(\log S)$ speed-up ratio over the conventional $O(n \log n)$ software implementation time, and since it is a consequence of the I/O consideration, this upper bound holds independently of device speed. Similar upper bounds have been established for speed-up ratios achievable by devices for other computations such as sorting and matrix multiplication [18, 42]. Knowing the I/O-imposed performance limit helps prevent overkill in the design of a special-purpose device.

In practice, problems are typically "larger" than special-purpose devices. Therefore, questions such as how a computation can be decomposed to minimize I/O, how the required I/O time is related to the size of a special-purpose system and its memory, and how the I/O bandwidth limits the speed-up ratio achievable by a special-purpose system present another set of challenges to the system architect.

75

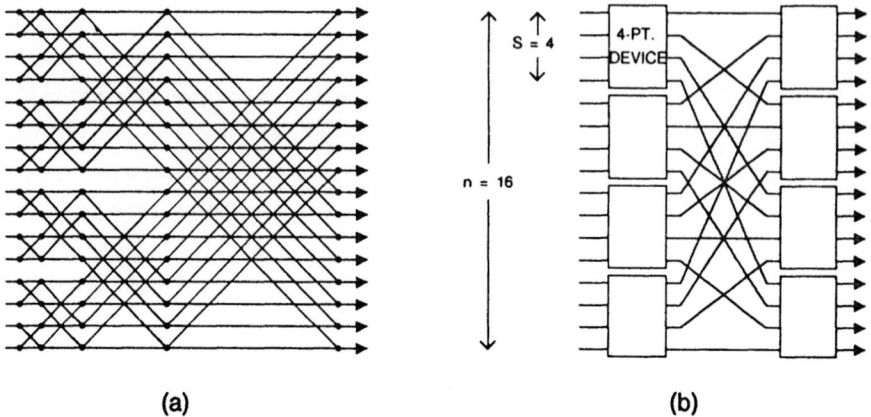

Figure 2-1. (a) 16-point FFT graph; (b) decomposing the FFT computation with $n = 16$ and $S = 4$.

2.2 Systolic Architectures: the Basic Principle

As a solution to the above challenges, we consider systolic architectures, an architectural concept originally proposed for VLSI implementation of some matrix operations [26]. Examples of systolic architectures follow in the next lecture, which contains a walk-through of a family of designs for the convolution computation.

A systolic system consists of a set of interconnected *cells*, each capable of performing some simple operation. Because simple, regular communication and control structures have substantial advantages over complicated ones in design and implementation, cells in a systolic system are typically interconnected to form a systolic array or a systolic tree. Information in a systolic system flows between cells in a pipelined fashion, and communication with the outside world occurs only at the "boundary cells." For example, in a systolic array, only those cells on array boundaries may be I/O ports for the system.

Computational tasks can be conceptually classified into two families — compute-bound computations and I/O-bound computations. In a computation, if the total number of operations is larger than the total number of input and output elements, then the computation is compute-bound, otherwise it is I/O-bound. For example, the ordinary matrix-matrix multiplication algorithm represents a compute-bound task, since every entry in a matrix is multiplied by all entries in some row or column of the other matrix. Adding two matrices, on the other hand, is I/O-bound, since the total number of adds is not larger than the total number of entries in the two matrices. It should be clear that any attempt to speed up an I/O-bound computation must rely on an increase in memory bandwidth. Memory bandwidth can be increased by the use of either fast components (which could be expensive) or interleaved memories (which could create complicated memory management problems). Speeding up a compute-bound computation, however, may often be accomplished in a relatively simple and inexpensive manner, that is, by the systolic approach.

The basic principle of a systolic architecture, a systolic array in particular, is illustrated in Figure 2-2. By replacing a single processing element with an array of PEs, or cells in the terminology of this article, a higher computation throughput can be achieved without increasing memory bandwidth. The function of the memory in the diagram is analogous to that of the heart; it "pulses" data (instead of blood) through the array of cells. The crux of this approach is to ensure that once a data item is brought out from the memory it can be used effectively at each cell it passes while being "pumped" from cell to cell along the array. This is possible for a wide class of compute-bound computations where multiple operations are performed on each data item in a repetitive manner.

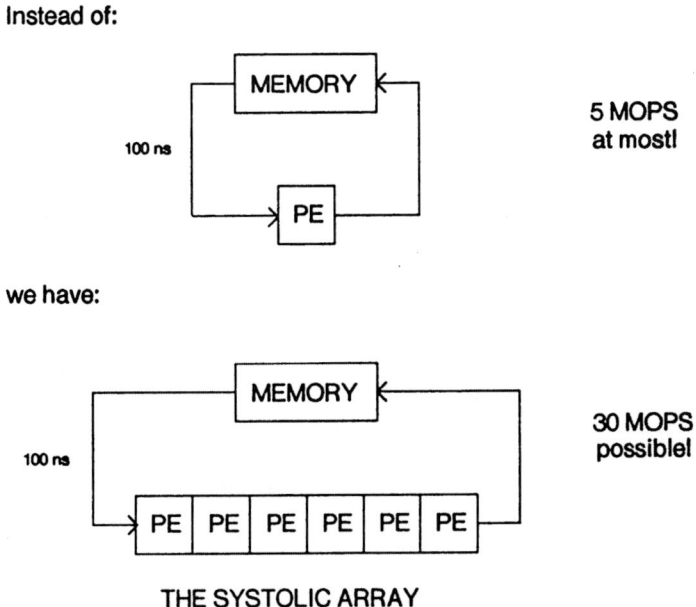

Figure 2-2. Basic principle of a systolic system.

Being able to use each input data item a number of times (and thus achieving high computation throughput with only modest memory bandwidth) is just one of many advantages of the systolic approach. Other advantages, such as modular expansibility, simple and regular data and control flows, use of simple and uniform cells, elimination of global broadcasting and fan-in, and (possibly) fast response time, will be illustrated in systolic designs in the next lecture.

2.3 Systolic Designs

Systolic designs using (one- or two-dimensional) array or tree structures are available for the following regular, compute-bound computations.

- Signal and image processing:

    o FIR and IIR filtering, and 1-D convolution [20, 21, 8];

    o 2-D convolution and correlation [28, 4, 49, 27, 22];

    o discrete Fourier transform [20, 21];

    o interpolation [27];

    o 1-D and 2-D median filtering [11]; and

    o geometric warping [27].

- Matrix arithmetic:

    o matrix-vector multiplication [26];

    o matrix-matrix multiplication [26, 47];

    o matrix triangularization (solution of linear systems, matrix inversion) [26, 15];

    o QR-decomposition (eigenvalue, least squares computations) [5, 15]; and

    o solution of triangular.linear systems [26].

- Non-numeric applications:

    o data structures — stack and queue [16], searching [42, 3, 38], priority queue [30], and sorting [30, 42];

    o graph algorithms — transitive closure [17] minimum spanning trees [2], and connected components [40];

    o geometric algorithms — convext hull generation [9];

    o language recognition — string matching [13] and regular expression [14];

    o dynamic programming [17];

    o polynomial algorithms — polynomial mulltiplication and division [23], and polynomail greatest common divisors [6];

o relational data-base operationse [25, 29]; and

o Monte Carlo Simulation [48].

In general, systolic designs apply to any compute-bound problem that is regular — that is, one where repetitive computations are performed on a large set of data. Thus, the above list is certainly not complete (and was not intended to be so). Its purpose is to provide a range of typical examples and possible applications. After studying several of these examples, one should be able to start designing systolic systems for one's own tasks; some systolic solution can usually be found without too much difficulty. This is probably due to the fact that most compute-bound problems are inherently regular in the sense that they are definable in terms of simple recurrences. Indeed, the notion of systolicity is implicit in quite a few previously known special-purpose designs, such as the sorting [46] and multiplier designs. [33] This should not come as a surprise; as we have been arguing, systolic structures are essential for obtaining any cost-effective, high-performance solution to compute-bound problems. It is useful, however, to make the systolic concept explicit so that designers will be conscious of this important design criterion.

3. Systolic Arrays for Convolution-Like Computations

In this section we will systematically derive a family of systolic arrays for the convolution problem. These designs can be generalized to other problems, including signal resampling, pattern matching, regular language recognition, polynomial evaluation (including discrete Fourier transform), polynomial multiplication and division, nearest neighbor searching, convex hull generation, recursive filtering, Monte Carlo simulation, etc.

3.1 The Convolution Computation

To provide concrete examples of various systolic structures, we present a family of systolic designs for the convolution problem, which is defined as follows:

Given the sequence of weights $\{w_1, w_2, \ldots, w_k\}$, and the input sequence $\{x_1, x_2, \ldots, x_n\}$,

compute the result sequence $\{y_1, y_2, \ldots, y_{n+1-k}\}$ defined by

$$y_i = w_1 x_i + w_2 x_{i+1} + \ldots + w_k x_{i+k-1}.$$

We consider the convolution problem because it is a simple problem with a variety of enlightening systolic solutions, because it is an important problem in its own right, and more importantly, because it is representive of a wide class of computations suited to systolic designs. The convolution problem, also called the FIR-filtering problem or the correlation problem, can be viewed as a problem of combining two data-streams, w_i's and x_i's, in a certain manner (as in the above equation) to form a resultant data-stream of y_i's. This type of computation is common to a wide class of computation routines. For example, if multiplication and addition are interpreted as comparison and boolean AND, respectively, then the convolution problem becomes the pattern matching problem [13]. Architectural concepts for the convolution problem can thus be applied to these other problems as well.

The convolution problem is compute-bound, since each input x_i is to be multiplied by each of the k weights. If the x_i is input separately from memory for each multiplication, then when k is large, memory bandwidth becomes a bottleneck, precluding a high-performance solution. As indicated earlier, a systolic architecture resolves this I/O bottleneck by making multiple use of each x_i fetched from the memory. (The conceptual schema in Figure 3-1 depicts communication between a systolic convolution array and the memory. Result y_i's could be output from the systolic array at either of its end cells.) Based on this principle, several systolic designs for solving the convolution problem are described below. For simplicity, all illustrations assume that $k = 3$.

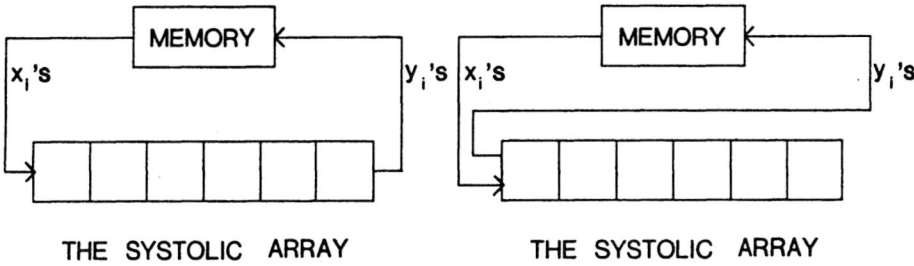

Figure 3-1. Communication between a systolic convolution array and the memory. Results can be output from either of the array's end cells as shown in (a) and (b).

3.2 (Semi-) Systolic Convolution Arrays with Global Data Communication

If an x_i, once brought out from the memory, is broadcast to a number of cells, then the same x_i can be used by all the cells. This broadcasting technique is probably one of the most obvious ways to make multiple use of each input element. The opposite of broadcasting is *fan-in*, through which data items from a number of cells can be collected. The fan-in technique can also be used in a straightforward manner to resolve the I/O bottleneck problem. In the following, we describe systolic designs that utilize broadcasting and fan-in.

Design B1 — *Broadcast inputs, move results, weights stay*

The systolic array and its cell definition are depicted in Figure 3-2. Weights are preloaded to the cells, one at each cell, and stay at the cells throughout the computation. Partial results y_i move systolically from cell to cell in the left-to-right direction, that is, each of them moves over the cell to its right during each cycle. At the beginning of a cycle, one x_i is broadcast to all the cells and one y_i, initialized as zero, enters the left-most cell. During cycle one, $w_1 x_1$ is accumulated to y_1 at the left-most cell, and during cycle two, $w_1 x_2$ and $w_2 x_2$ are accumulated to y_2 and y_1 at the left-most and middle cells, respectively. Starting from cycle three, the final (and correct) values of $y_1, y_2, \ldots$ are output from the right-most cell at the rate of one y_i per cycle. The basic principle of this design was previously proposed for circuits to implement a pattern matching processor [36] and for circuits to implement polynomial multiplication [10, 19, 32, 39].

Design B2 — *Broadcast inputs, move weights, results stay*

In design B2 (see Figure 3-3) each y_i stays at a cell to accumulate its terms, allowing efficient use of available multiplier-accumulator hardware. (Indeed, this design is described in an application booklet for the

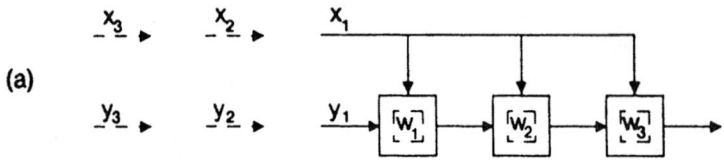

(a)

(b)

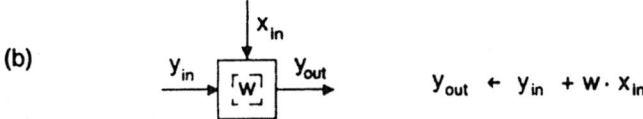

$$y_{out} \leftarrow y_{in} + w \cdot x_{in}$$

Figure 3-2. Design B1: systolic convolution array (a) and cell (b) where x_i's are broadcast, w_i's stay, and y_i's move systolically.

TRW multiplier-accumulator chips [41].) The weights circulate around the array of cells, and the first weight w_1 is associated with a tag bit that signals the accumulator to output and resets its contents.* In design B1 (Figure 3-2), the systolic path for moving y_i's may be considerably wider than that for moving w_i's in design B2 because for numerical accuracy y_i's typically carry more bits than w_i's. The use of multiplier-accumulators in design B2 may also help increase precision of the results, since extra bits can be kept in these accumulators with modest cost. Design B1, however, does have the advantage of not requiring a separate bus (or other global network), denoted by a dashed line in Figure 3-3, for collecting outputs from individual cells.

Design F — *Fan-in results, move inputs, weights stay*

If we consider the vector of weights (w_k, w_{k-1}, ..., w_1) as being fixed in space and input vector (x_n, x_{n-1}, ..., x_1) as sliding over the weights in the left-to-right direction, then the convolution problem is one that computes the inner product of the weight vector and the section of input vector it overlaps. This view suggests the systolic array shown in Figure 3-4. Weights are preloaded to the cells and stay there throughout the computation. During a cycle, all the x_i's move one cell to the right, multiplications are performed at all cells simultaneously, and their results are fanned-in and summed using an adder to form a new y_i. When the number of cells, k, is large, the adder can be implemented as a pipelined adder tree to avoid large delays in each cycle. Designs of this type using unbounded fan-in have been known for quite a long time, for example, in signal processing [44], in coding networks [39], and in pattern matching [35].

---

*To avoid complicated pictures, control structures such as the use of tag bits to gate outputs from cells are omitted from the diagrams of this article.

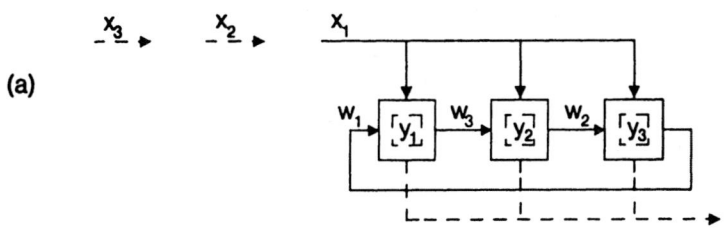

(a)

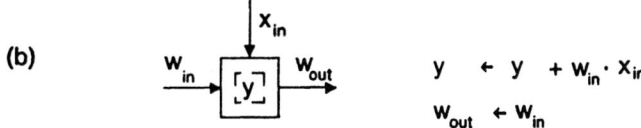

(b)

$$y \leftarrow y + w_{in} \cdot x_{in}$$

$$w_{out} \leftarrow w_{in}$$

Figure 3-3. Design B2: systolic convolution array (a) and cell (b) where x_i's are broadcast, y_i's stay, and w_i's move systolically.

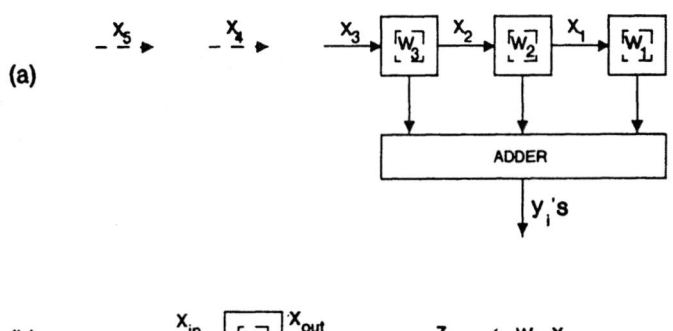

(a)

(b)

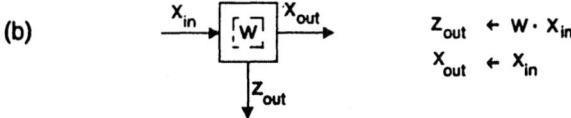

$$z_{out} \leftarrow w \cdot x_{in}$$

$$x_{out} \leftarrow x_{in}$$

Figure 3-4. Design F: systolic convolution array (a) and cell (b) where w_i's stay, x_i's move systolically, and y_i's are formed through the fan-in of results from all the cells.

3.3 (Pure-) Systolic Convolution Arrays without Global Data Communication

Although global broadcasting or fan-in solves the I/O bottleneck problem, implementing it in a modular, expandable way presents another problem. Providing (or collecting) a data item to (or from) all the cells of a systolic array, during each cycle, requires the use of a bus or some sort of tree-like network. As the number of cells increases, wires become long for either a bus or tree structure; expanding these non-local communication paths to meet the increasing load is difficult without slowing down the system clock. This engineering difficulty of extending global networks is significant at chip, board, and higher levels of a computer system. Fortunately, as will be demonstrated below, (pure-) systolic convolution arrays without global data communication do exist. Potentially, these arrays can be extended to include an arbitrarily large number of cells without encountering engineering difficulties [12].

Design R1 — *Results stay, inputs and weights move in opposite directions*

In design R1 (see Figure 3-5) each partial result y_i stays at a cell to accumulate its terms. The x_i's and w_i's move systolically in opposite directions such that when an x meets a w at a cell, they are multiplied and the resulting product is accumulated to the y staying at that cell. To ensure that each x_i is able to meet every w_i, consecutive x_i's on the x datastream are separated by two cycle times and so are the w_i's on the w datastream.

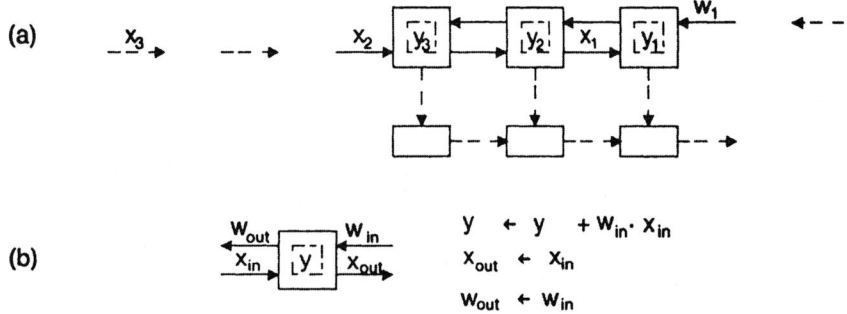

Figure 3-5. Design R1: systolic convolution array (a) and cell (b) where y_i's stay and x_i's and w_i's move in opposite directions systolically.

Like design B2, design R1 can make efficient use of available multiplier-accumulator hardware; it can also use a tag bit associated with w_1 to trigger the output and reset the accumulator contents of a cell. Design R1 has the advantage that it does not require a bus, or any other global network, for collecting output from cells; a *systolic output path* (indicated by broken arrows in Figure 3-5) is sufficient. Because consecutive w_i's are well separated (by two cycle times), a potential conflict — that more than one y_i may reach a single latch on the systolic output path simultaneously — cannot occur. It can also be easily checked that the y_i's will output from the systolic output path in the natural ordering, $y_1, y_2, \ldots$. The basic idea of this design, including that of the systolic output path, has been used to implement a pattern matching chip [13].

Notice that in Figure 3-5 only about one-half the cells are doing useful work at any time. To fully utilize the potential throughput, two independent convolution computations can be interleaved in the same systolic array, but cells in the array would have to be modified slightly to support the interleaved computation. For example, an additional accumulator would be required at each cell to hold a temporary result for the other convolution computation.

Design R2 — *Results stay, inputs and weights move in the same direction but at different speeds*

One version of design R2 is illustrated in Figure 3-6. In this case both the x and w datastreams move from left to right systolically, but the x_i's move twice as fast as the w_i's. More precisely, each w_i stays inside every cell it passes for one extra cycle, thus taking twice as long to move through the array as any x_i. In this design, multiplier-accumulator hardware can be used effectively, and so can the tag bit method to signal the output of the accumulator contents at each cell. Compared to design R1, this design has the advantage that all cells work all the time when performing a single convolution, but requires an additional register in each cell to hold a w value. This algorithm has been used for implementing a pipeline multiplier [33].

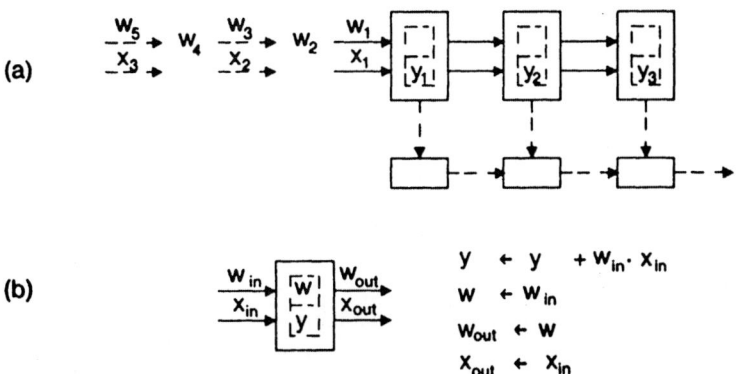

Figure 3-6. Design R2: systolic convolution array (a) and cell (b) where y_i's stay and x_i's and w_i's both move systolically in the same direction but at different speeds.

There is a dual version of design R2; we can have the w_i's move twice as fast as the x_i's. To create delays for the x data-stream, this dual design requires a register in each cell for storing an x rather than a w value. For circumstances where the w_i's carry more bits than the x_i's, the dual design becomes attractive.

Design W1 — *Weights stay, inputs and results move in opposite directions*

In design W1 (and design W2, below), weights stay, one at each cell, but results and inputs move systolically. These designs are not geared to the most effective use of available multiplier-accumulator hardware, but for some other circumstances they are potentially more efficient than the other designs. Because the same set of weights is used for computing all the y_i's and different sets of the x_i's are used for computing different y_i's, it is natural to have the w_i's preloaded to the cells and stay there, and let the x_i's and the y_i's move along the array. We will see some advantages of this arrangement in the systolic array depicted in Figure 3-7, which is a special case of a proposed systolic filtering array [21]. This design is fundamental in the sense that it can be naturally extended to perform recursive filtering [20, 21] and polynomial division [23].

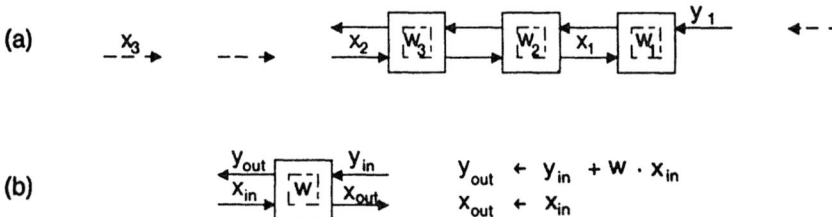

Figure 3-7. Design W1: systolic convolution array (a) and cell (b) where w_i's stay and x_i's and y_i's move systolically in opposite directions.

In design W1, the w_i's stay and the x_i's and y_i's move systolically in opposite directions. Similar to design R1, consecutive x_i's and y_i's are separated by two cycle times. Note that because the systolic path for moving the y_i's already exists, there is no need for another systolic output path as in designs R1 and R2. Furthermore, for each i, y_i outputs from the left-most cell during the same cycle as its last input, x_{i+k-1} (or x_{i+2} for $k = 3$), enters that cell. Thus, this systolic array is capable of outputting a y_i every two cycle times with *constant response time*. Design W1, however, suffers from the same drawback as design R1, namely, only approximately one-half the cells work at any given time unless two independent convolution computations are interleaved in the same array. The next design, like design R2, overcomes this shortcoming by having both the x_i's and y_i's move in the same direction but at two speeds.

Design W2 — *Weights stay, inputs and results move in the same direction but at different speeds*

With design W2 (Figure 3-8), all the cells work all the time, but it loses one advantage of design W1, the constant response time. The output of y_i now takes place k cycles after the last of its inputs starts entering the left-most cell of the systolic array. This design has been extended to implement 2-D convolutions [28, 27], where high throughputs rather than fast responses are of concern. Similar to design R1, design W2 has a dual version for which the x_i's move twice as fast as the y_i's.

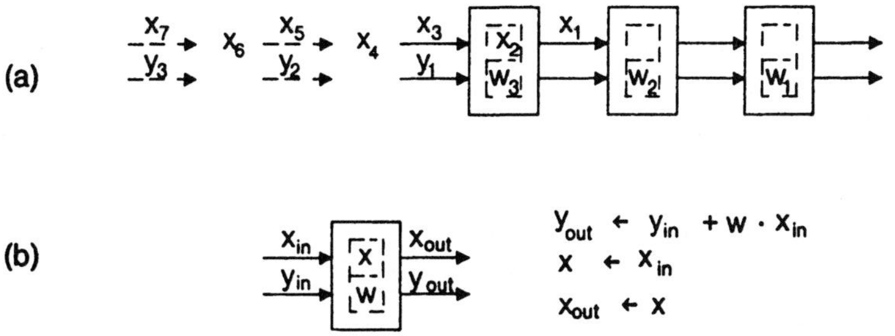

(a)

(b)

Figure 3-8. Design W2: systolic convolution array (a) and cell (b) where w_i's stay and x_i's and y_i's both move systolically in the same direction but at different speeds.

3.4 Remarks

The designs presented above by no means exhaust all the possible systolic designs for the convolution problem. For example, it is possible to have systolic designs where results, weights, and inputs all move during each cycle. It could also be advantageous to include inside each cell a "cell memory" capable of storing a set of weights. With this feature, using a *systolic address* (or *control*) *path* weights can be selected on-the-fly to implement interpolation, signal resampling or adaptive filtering [27]. Moreover, the flexibility introduced by the cell memories and systolic controls can make the same systolic array implement different functions. Indeed, the ESL systolic processor [4, 49] utilizes cell memories to implement multiple functions including convolution and matrix multiplication.

Once one systolic design is obtained for a problem, it is likely that a set of other systolic designs can be derived similarly. The challenge is to understand precisely the strengths and drawbacks of each design so that an appropriate design can be selected for a given environment. For example, if there are more weights than cells, it's useful to know that a scheme where partial results stay generally requires less I/O than one where partial results move, since the latter scheme requires partial results to be input and output many times. A single multiplier-accumulator hardware component often represents a cost-effective implementation of the multiplier and adder needed by each cell of a systolic convolution array. However, for improving throughput, sometimes it may be worthwhile to implement multiplier and adder separately to allow overlapping of their executions. Figure 3-9 depicts such a modification to design W1. Similar modifications can be made to other systolic convolution arrays. Another interesting scenario is the following one. Suppose that one or several cells are to be implemented directly with a single chip and the chip pin bandwidth is the implementation bottleneck. Then, since the basic cell of some semi-systolic convolution array such as design B1 or F requires only three I/O ports, while that of a pure-systolic convolution array always requires four, a semi-systolic array may be preferable in this case for saving pins, despite the fact that it requires global communication.

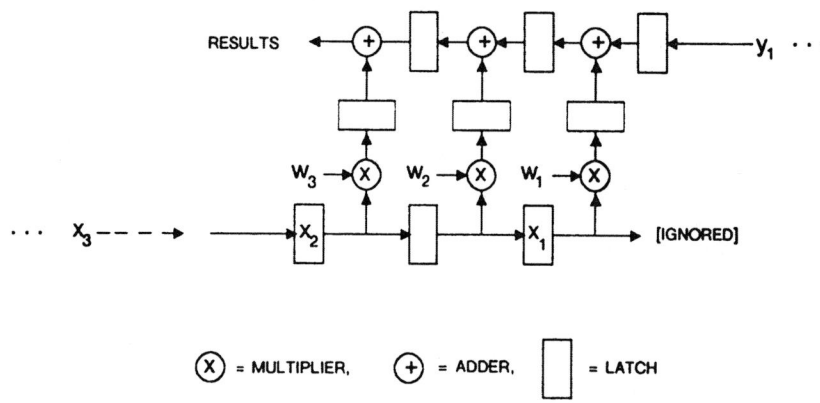

Figure 3-9. Overlapping the executions of multiply and add in design W1.

While numerous systolic designs are known today, the question of their automatic design is still open in general. However, designs presented in this lecture can be generalized in a straightforward way to other convolution-like computations such as signal resampling, pattern matching, regular language recognition, polynomial evaluation (including discrete Fourier transform), polynomial multiplication and division, nearest neighbor searching, convex hull generation, recursive filtering, Monte Carlo simulation, etc. Moreover, recent efforts show significant progress on more formal ways of converting non-systolic designs into systolic ones [47, 31]. Leiserson and Saxe, for instance, can convert some semi-systolic systems involving broadcasting or unbounded fan-in into pure-systolic systems without global data communication [31]. A related open research issue concerns the specification and verification of systolic structures. For implementation and proof purposes, rigorous notation other than informal pictures (as used in these notes) for specifying systolic designs is desirable.

4. PSC: A Programmable Systolic Chip

We have seen from the previous sections that systolic architectures are well suited for many high-performance, special-purpose systems. For well-understood, highly important inner loops a systolic processor could be cost-effectively built for each single application. For applications where more flexibility is desired, however, multi-purpose systolic arrays could be more attractive. An example is the ESL systolic processor, a linearly connected array of 28 cells, which can be microcoded to perform 2-D convolution, matrix multiplication, and other similar operations [4, 49]. Figure 4-1 depicts the ESL systolic processor, which has been running since January 1982, and can be called as a subroutine by FORTRAN programs running on a host VAX-11/780.

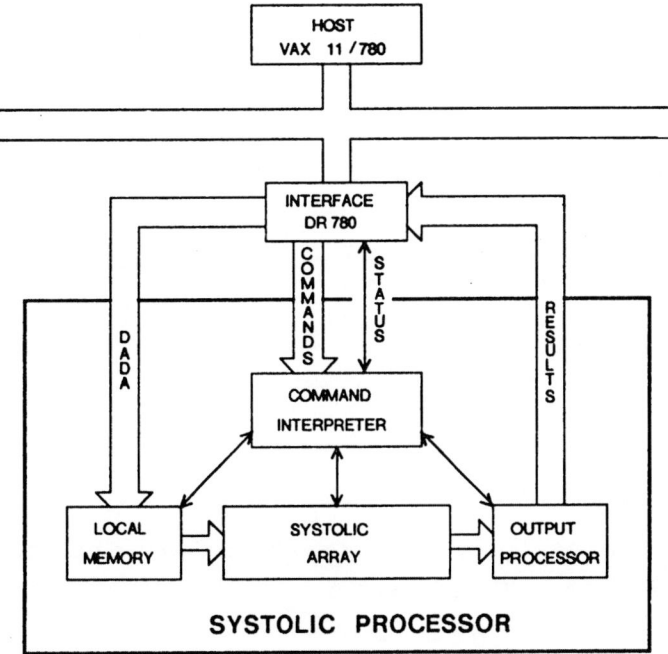

Figure 4-1. The ESL systolic processor.

The highest degree of flexibility can be obtained by allowing programmability in cells as well as reconfigurability of cell interconnections. The *programmable systolic chip* (PSC), which is nearly ready for fabrication in NMOS, is a prototype single-chip processor aimed at exploring the design space of systolic processors with this highest degree of flexibility. As illustrated in Figure 4-2, the chip will form the basic cells for a wide variety of systolic arrays. Note that conventional, commercially available microprocessor components are not adequate for forming cells in systolic arrays, because they do not have enough I/O bandwidth to pass data from cell to cell with a speed sufficient to balance the computation speed. Moreover, unlike the PSC, conventional microprocessors do not have fast, on-chip multiplier circuits which are crucial for high-speed signal and image processing.

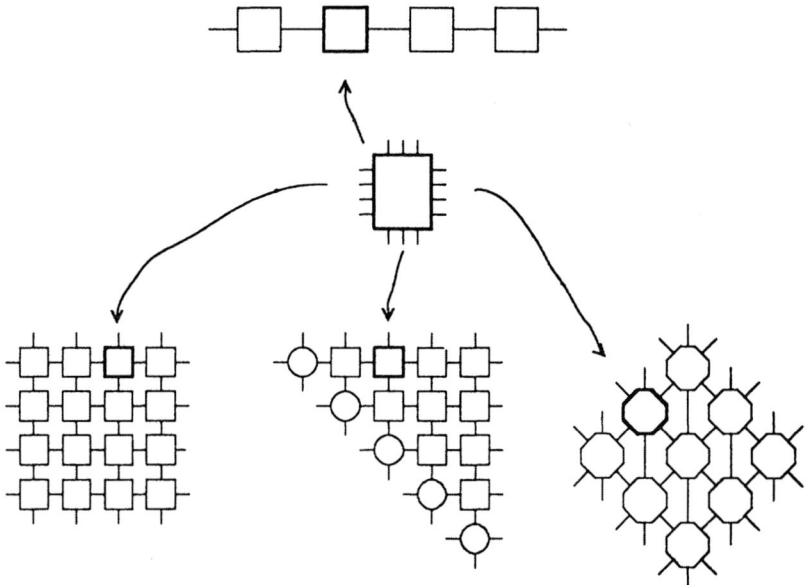

Figure 4-2. PSC: A building-block chip for a variety of systolic arrays.

By using arrays consisting of tens or hundreds of copies of the PSC chip, we expect to get orders of magnitude improvements in speed for applications including:

- Convolution-like operations in signal and image processing.

- Reed-Solomon encoding and decoding for error corrections.

- Encryption and finger-print calculations.

- Number theoretic transforms.

- Disk sorting.

Consider the application to Reed-Solomon error-correcting codes as an example. Suppose that each codeword consists of 224 information symbols followed by 32 check symbols and each symbol is an 8-bit integer; using the Reed-Solomon scheme, errors involving no more than 16 symbols can be corrected. We estimate that by using a linear array of 112 PSC chips the Reed-Solomon decoding can be performed in a throughput of 10 million bits per second. Encoding is much easier; it requires only about 30 PSC chips to achieve the same throughput. As far as we know, the fastest existing Reed-Solomon decoder with the same characteristics uses about 500 chips but achieves a throughput of only 1 million bits per second. The performance of the PSC-based system is largely due to the use of systolic array; for example, a highly effective systolic array has been developed for implementing the extended Euclidean computation, the most difficult task in the decoder implementation [6].

We see that there is a spectrum of possibilities for systolic processor designs, ranging from single-purpose processors to general-purpose, programmable processors such as the PSC chip, as depicted conceptually in Figure 4-3. The choice of a particular architecture in the design space depends on each individual application problem at hand.

It is worthwhile to note that a systolic processor has recently been implemented by the Naval Ocean Systems Center in San Diego as a testbed for a variety of systolic array configurations, and to understand the architectural features needed by various systolic algorithm [7, 45]. This processor has an 8×8 array of cells, each containing a Intel Arithmetic Processing Unit and microprocessor for achieving the flexibility needed in the testbed.

This chip is designed with all these applications in mind and its architecture is driven by program needs. That is, we develop programs (at micro-code level) first before we finalize the architecture. Features of the chip, as depicted in Figure 4-4, are summarized as follows::

- 3 eight-bit data input ports and 3 eight-bit data output ports

- 3 one-bit control input ports and 3 one-bit control output ports

- Eight-bit ALU with support for multiple precision and modulo 257 arithmetic

- Multiplier-accumulator (MAC) with eight-bit operants and 16-bit accumulator

- 64-word by 60-bit writable control store

- 64-word by 9-bit register file

- Three 9-bit on-chip buses

FLEXIBILITY

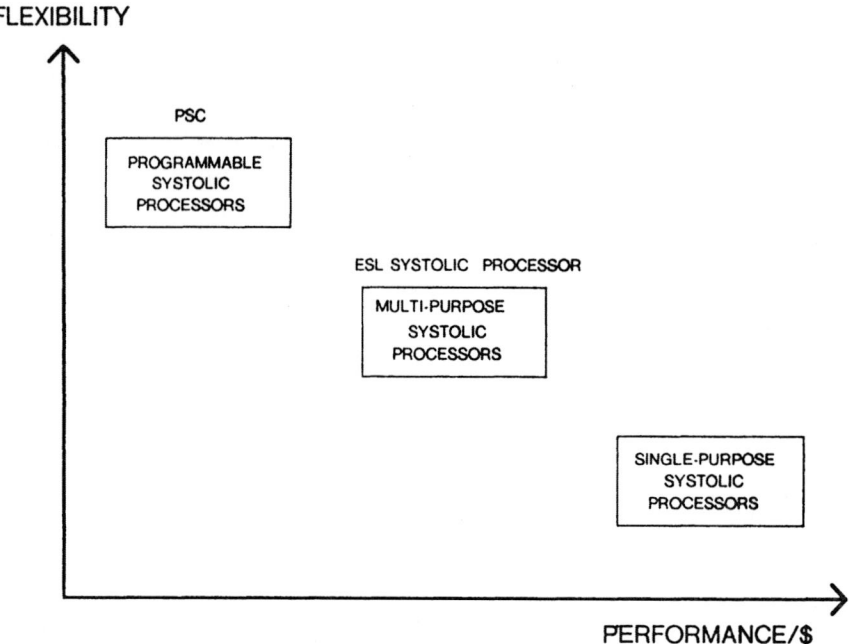

Figure 4-3. Design space for systolic processors.

- Stack-based microsequencer

The initial design of the PSC was specified in ISPS [1], and microcode for several systolic algorithms was written and tested as an aid to design evaluation. This process was iterated several times as experiments and measurements resulted in changes to the specification. Concurrently, design and layout of several individual pieces of the chip were begun. At a later stage, estimates of chip area contributed to the refinement of the processor's design, which was finalized in May of 1982. Currently, all of the major computational elements of the processor have been laid out, and only the final integration of the chip remains to be done.

The first version of the PSC is to be implemented in NMOS using Mead-Conway design rules [34] with λ = 2 microns. The chip contains about 20,000 transistors and 72 contact pads, and will be about 30 mm^2. Spice simulations indicate that the chip should operate at a clock rate between 3 and 4 MHz. The layout of the chip will be completed by mid-October, and chips packaged in 84 contact leadless chip carriers should be available for testing by the end of the year.

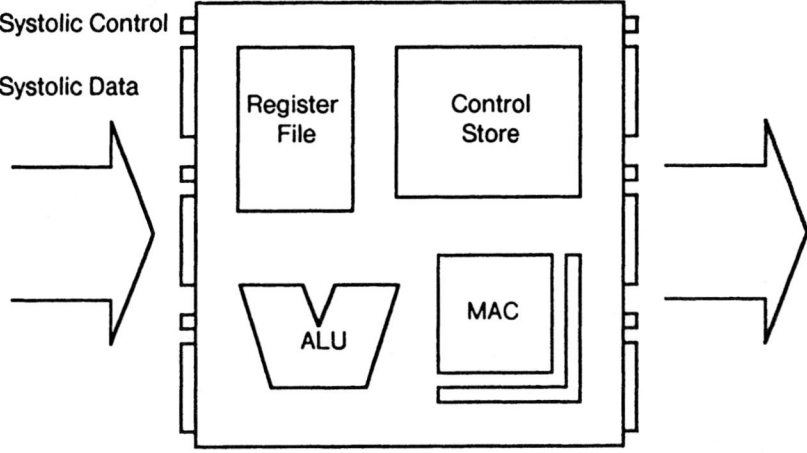

Figure 4-4. PSC Features

5. System Integration Problem

With the development of systolic architectures, more and more special-purpose systems may become feasible — especially systems that implement fixed, well-understood computation routines. This together with other advances, such as fast FFT machines and chips that are becoming commercially available, we feel that design and implementation costs of many special-purpose devices are no longer a major difficulty. A pressing problem now is that of system issues such as convenient incorporation of these devices into complete systems, and their effective utilization from a system point of view. This section will discuss some aspects of the problem.

5.1 System Issues Illustrated by Examples

We identify relevant system issues by studying some concrete examples. To illustrate the issues it suffices to consider the examples separately, although in practice several of them may actually be "chained" together in one single application. For instance, some signal processing applications require that a matrix multiplication be performed following an FFT computation.

5.1.1 Matrix Multiplication

Suppose that we want to build a system that can incorporate special-purpose devices for performing matrix multiplications. Any of such devices at the hardware level can only multiply matrices of some fixed orders. Therefore for a given problem of arbitrary size, we often must decompose it into subproblems so that each of these subproblems can be solved by one of these special-purpose devices. In the following we assume that the given matrices are $A = (a_{ij})$ and $B = (b_{ij})$, both $n \times n$, and that we want to compute their product $C = (c_{ij})$. For illustration the only special-purpose device that we will consider is a straightforward one-dimensional systolic array described below.

One-dimensional systolic array for matrix multiplication. The array consists of k cells, each capable of performing a multiply-accumulation operation, as depicted in Figure 5-1. We assume that k is a constant much less than n. For $j = 1, 2, \ldots, n$, the j-th cell from the left, computes the inner product of the two vectors $(a_{i,1}, a_{i,2}, \ldots, a_{i,n})$ and $(b_{1,j}, b_{2,j}, \ldots, b_{n,j})$. Thus the whole array computes the product of the row vector $(a_{i,1}, a_{i,2}, \ldots, a_{i,n})$ and the matrix B' consisting of the first k columns of the matrix B. By pumping all the row vectors in A into the array one after another and by recirculating the $(b_{1,j}, b_{2,j}, \ldots, b_{n,j})$ around cell j for each j, matrix multiplication $C' = A \times B'$ is performed.

Three levels of I/O requirements. Since the systolic array above can handle k columns of matrix B in one pass, to use it to multiply A and B it is natural to decompose B into submatrices $B_1, B_2, \ldots, B_h$, where $h = \lceil n/k \rceil$, each B_j has k columns of B for $j = 1, \ldots, h-1$, and B_h may have less than k columns of B. The product

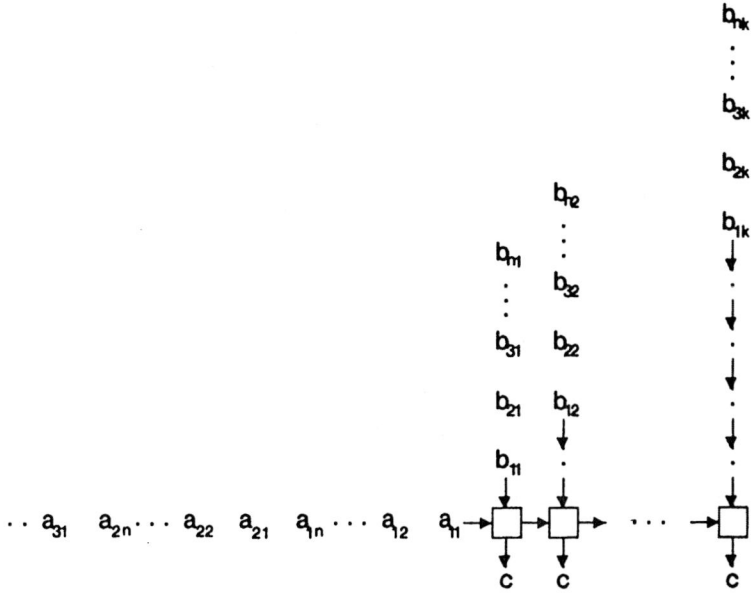

Figure 5-1. One-dimensional systolic matrix multiplication array.

C of A and B is decomposed similarly into submatrices $C_1, C_2, \ldots, C_h$. The systolic array computes $C_1 = A \times B_1$ first, $C_2 = A \times B_2$ next, and so on. One can check that each element in A enters the systolic array h times, whereas each element in B enters the array n times. Elements in C never have to be input to the array, and they are output only after their final values have been accumulated. Thus the I/O requirement for B is the highest, that for A is the second and that for C is the lowest. This suggests the use of hierarchical memories with different sizes and access rates to store the three matrices. A challenging problem is to figure out a near-optimal configuration of memories that will suit well to a large class of devices and problem sizes.

5.1 .2 2-D Convolution

One of the most compute-intensive tasks in image processing is the two-dimensional (2-D) convolution problem. Consider the problem of convolving an $m \times m$ kernel with an $n \times n$ image, where $m \leq n$. Assume that m is not large enought to make an FFT-based solution cost-effective. Then the straightforward method of solving the problem will require $O(m^2 n^2)$ operations. Several systolic devices for the 2-D convolution problem have been recently proposed (see, e. g., [22, 27, 49]).

Problem decomposition. Suppose that such a device can only handle kernels of size $k \times k$, where $k \leq m$. Then the $m \times m$ kernel has to be decomposed into subkernels of size $k \times k$. Consider the algorithm that slides the kernel from left to right until the right boundary of the image is researched. Then slide down one row of image and do the same, until the bottom of the image is researched. For the subkernels, first input subkernel $K_{1,1}$ to the 2-D convolution device, and then pass the first k rows of image through the device; some partial sums are obtained. Do the same for subkernels: $K_{1,2}, \ldots, K_{1, \lceil m/k \rceil}$ and add the resulting partial sums together. Now input subkernel $K_{2,1}$ to the 2-D convolution device, and pass the second k rows of the image through the device and add the resulting partial sums. The algorithm is illustrated in figure 5-2.

I/O requirement. From this algorithm, it is easy to see that kernel elements have to enter the device $O(n)$ times, and entries of image $O(m^2/k)$ times. Thus the I/O requirement for the kernel is the highest, that for the input image is the second, and that for the resulting image is the lowest.

5.1 .3 FFT

Consider the problem of computing the n-point discrete Fourier transform (DFT) by the fast Fourier transform (FFT) algorithm. Suppose that we have a special-purpose device that can compute k-point DFT's, where k is much smaller than n.

Problem decomposition and I/O Requirements. Decomposition for the FFT was discussed earlier in Lecture 1 (see Figure 2-1). We note that the I/O requirement for the FFT computation is inversely proportional to the logarithm of the memory size. Thus it is possible to trade the size of a memory for its speed. Figure 5-3 (a) depicts a scenario that a special-purpose device for the FFT computation is supported by a two-level memory; Figure 5-3 (b) shows some of the memory size and speed trade-off results.

Control requirements. A fairly nontrivial task in the use of special-purpose devices for the FFT computation is the generation of subcomputation blocks as shown in Figure 2-1 (b). Inputs to such a block in general are outputs from other blocks multiplied by some so-called "twiddle" factors. This requires shuffling bits in addresses and generating the twiddle factors. Some custom hardware is likely needed here to keep up the speed.

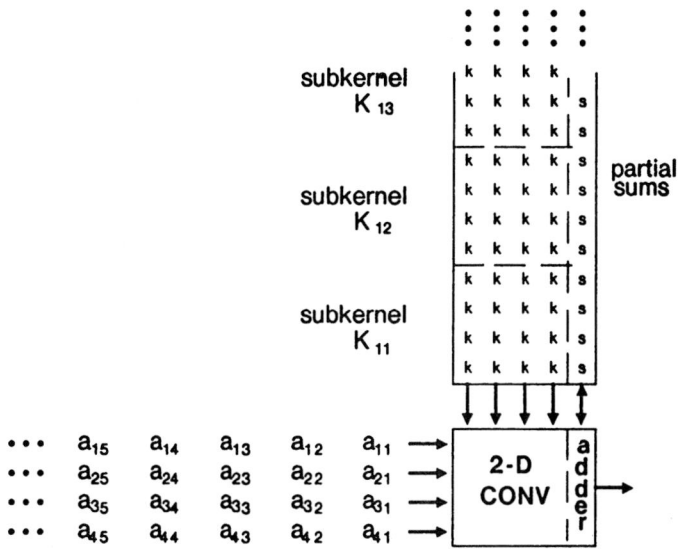

Figure 5-2. systolic 2-D convolution device

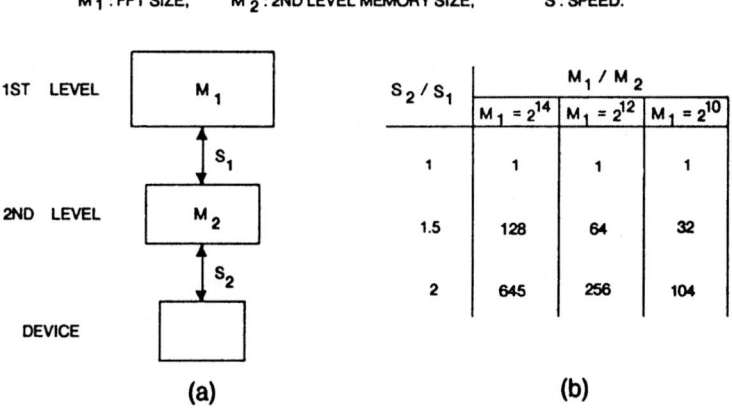

(a)

(b)

Figure 5-3. (a) two-level memory for the FFT computation; (b) memory size and speed trade-offs.

5.2 A Conceptual System

After having identified some of the issues for incorporating special-purpose devices in a system, here we outline a conceptual system addressing these issues.

5.2 .1 System Block Diagram

The system block diagram is shown in Figure 5-4.

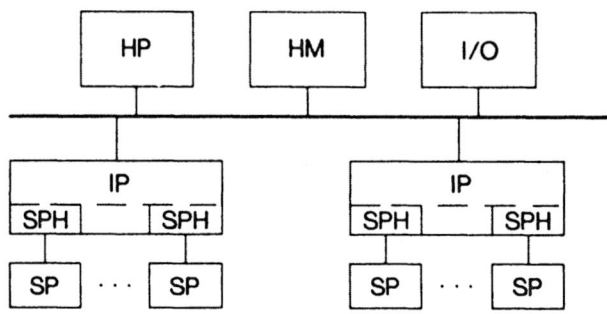

Figure 5-4. System block diagram.

Functions of the blocks are described in the following:

Host Processor — HP. This is the central controller for the system. It runs the operating system, and schedules and monitors activities of all components in the system. In the simplest form, it could just be a micro-store, loadable with microcodes from the outside world. But in some applications it may be necessary for the HP to perform some scalar operations that can not be done cost-effectively elsewhere including any special-purpose device in the system.

Host Memory — HM. This is the system memory that in theory is capable of holding all the input and output data for a given problem. For example, for the matrix multiplication problem it should be able to hold the two input matrices and the resulting matrix. In practice, for exceedingly large problems HM should at least be large enough to serve as a buffer so that communication with the outside world will not become a performance bottleneck for the system.

Host I/O — I/O. Through the I/O port, data and microcodes can be input to and out from the system.

Intermediate Processor — IP. This processor serves as the interface between the system and special-purpose devices. It has a relatively large (maybe interleaved) memory to provide a level of memory in the memory hierarchy called for by many applications including ones discussed in Section 2. Potentially, it can also coordinate those special-purpose devices to which it interfaces to implement closely coupled functions such as the "chaining" operation.

Special-purpose Processor — SP. An SP is a special-purpose or systolic processor for performing some high-level functions such as matrix multiplication and FFT; it may have some private memory to store data that are most frequently accessed. To the system SP's are interchangeable; therefore it is important that the I/O behaviors of the SP's that we are interested are uniform. Fortunately, this is the case for most of the systolic processors [24].

SP Handler — SPH. The SPH is a microprogrammable processor; every SP in the system is served by one specially programmed SPH as its controller. In addition, addresses of inputs to and output from an SP are generated by its SPH. These addresses are sent to the IP memory to which the SP interfaces.

5.3 System Programming

We have introduced the major components in the system. This section describes how the system should be programmed to fulfill its objective of providing a convenient environment for the users to utilize special-purpose devices available in the system.

5.3 .1 Compilation Phase

During this phase source programs submitted by users are transformed into programs directly executable by special-purpose devices. Major stages of this phase are shown in Figure 5-5.

Decomposition stage. As mentioned in Section 2, to utilize a special-purpose device a problem of arbitrary size may have to be decomposed into subproblems of smaller sizes so that each of the subproblems can be solved directly by the special-purpose device. Outputs from the decomposition stage are a collection of subproblems, information on the possible data-dependency among them, and recommendation on the execution order for their solutions for minimizing the I/O cost. To define a subproblem one has to specify where its input data come from and where its outputs should be stored. It appears that in general different decomposition routines are needed for different problems or different special-purpose devices.

Scheduling stage. Based on the outputs from the decomposition stage, the system configuration, and system component status, subproblems are assigned to various SP's during this stage. Outputs of this stages form a sequence of instructions to the SP's.

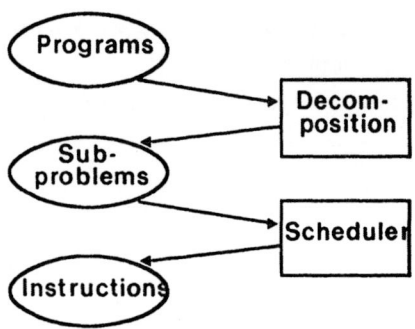

Figure 5-5. Stages in the compilation phase.

5.3 .2 Operation Phase

Codes produced during the compilation phase are carried out during the operation phase. The system is first set up by sending initialization parameters to its components. When operation phase starts, data required by an SP are sent from HM to the IP to which the SP interfaces. The SP retrieves its inputs from and stores its outputs to the IP memory according to the addresses generated by the associated SPH. The HP monitors activities at or above the IP level, and is responsible to data transfers among the IP's and the HM.

To understand system programming issues and the overall system control, at CMU we have written lisp programs to simulate systems that use special-purpose devices for performing matrix multiplication and FFT. Further simulations will be performed before the system architecture is finalized. The plan is to build a VAX-based prototype host machine with the concepts outlined above within a year.

References

[1]   Barbacci, M.R.
        Instruction Set Processor Specifications (ISPS): The Notation and Its Application.
        IEEE Transactions on Computers C-30(1):24-40, January, 1981.

[2]   Bentley, J.L.
        A Parallel Algorithm for Constructing Minimum Spanning Trees.
        Journal of Algorithms 1:51-59, 1980.

[3]   Bentley, J.L. and Kung, H.T.
        A Tree Machine for Searching Problems.
        In *Proceedings of 1979 International Conference on Parallel Processing*, pages 257-266. IEEE, August,
        1979.
        Also available as a CMU Computer Science Department technical report, August 1979.

[4]   Blackmer, J., P. Kuekes and Frank, G.
        A 200 MOPS Systolic Processor.
        In *Proceedings of SPIE Symposium, Vol. 298, Real-Time Signal Processing IV*. The Society of Photo-
        optical Instrumentation Engineers, August, 1981.

[5]   Bojanczyk, A., Brent, R.P. and Kung, H.T.
        Numerically Stable Solution of Dense Systems of Linear Equations Using Mesh-Connected Processors.
        Technical Report, Carnegie-Mellon University, Computer Science Department, May, 1981.
        The final version of the paper is to appear in *SIAM Journal on Scientific and Statistical Computing.*

[6]   Brent, R.P. and Kung, H.T.
        Systolic VLSI Arrays for Polynomial GCD Computation.
        Technical Report, Carnegie-Mellon University, Computer Science Department, May, 1982.

[7]   Bromley, K., Symanski, J.J., Speiser, J.M., and Whitehouse, H.J.
        Systolic Array Processor Developments.
        In Kung, H.T., Sproull, R.F., and Steele, G.L., Jr. (editors), *VLSI Systems and Computations*, pages
        273-284. Computer Science Department, Carnegie-Mellon University, Computer Science Press,
        Inc., October, 1981.

[8]   Cappello, P.R. and Steiglitz K.
        Digital Signal Processing Applications of Systolic Algorithms.
        In Kung, H.T., Sproull, R.F., and Steele, G.L., Jr. (editors), *VLSI Systems and Computations*, pages
        245-254. Computer Science Department, Carnegie-Mellon University, Computer Science Press,
        Inc., October, 1981.

[9]   Chazelle, Bernard.
        Computational Geometry on a Systolic Chip.
        Technical Report, Carnegie-Mellon University, Computer Science Department, May, 1982.

[10]   Cohen, D.
        Mathematical Approach to Computational Networks.
        Technical Report ISI/RR-78-73, University of Southern California, Information Sciences Institute,
        November, 1978.

[11]   Fisher, A.
        Systolic Algorithms for Running Order Statistics in Signal and Image Processing.
        In Kung, H.T., Sproull, R.F., and Steele, G.L., Jr. (editors), *VLSI Systems and Computations*, pages
            265-272. Computer Science Department, Carnegie-Mellon University, Computer Science Press,
            Inc., October, 1981.

[12]   Fisher, A.L. and Kung, H.T.
        Synchronizing Large Systolic Arrays.
        In *Proceedings of SPIE Symposium, Vol. 341, Real-Time Signal Processing V*. The Society of Photo-
            Optical Instrumentation Engineers, May, 1982.

[13]   Foster, M.J. and Kung, H.T.
        The Design of Special-Purpose VLSI Chips.
        Computer 13(1):26-40, January, 1980.
        Reprint of the paper appears in *Digital MOS Integrated Circuits*, edited by Elmasry, M.I., IEEE Press
            Selected Reprint Series, 1981, pp. 204-217. A preliminary version of the paper, entitled "Design of
            Special-Purpose VLSI Chips: Example and Opinions," also appears in *Proceedings of the 7th
            International Symposium on Computer Architecture*, pp. 300-307, La Baule, France, May 1980. .

[14]   Foster, M.J. and Kung, H.T.
        Recognize Regular Languages With Programmable Building-Blocks.
        In Gray, J.P. (editor), *VLSI 81*, pages 75-84. Academic Press, August, 1981.
        The final version is to appear in *Journal of Digital Systems*.

[15]   Gentleman, W.M. and Kung, H.T.
        Matrix Triangularization by Systolic Arrays.
        In *Proceedings of SPIE Symposium, Vol. 298, Real-Time Signal Processing IV*. The Society of Photo-
            optical Instrumentation Engineers, August, 1981.

[16]   Guibas, L.J. and Liang, F.M.
        Systolic Stacks, Queues, and Counters.
        In *Proceedings of the Conference on Advanced Research in VLSI*. Cambridge, Massachusetts, January,
            1982.

[17]   Guibas, L.J., Kung, H.T. and Thompson, C.D.
        Direct VLSI Implementation of Combinatorial Algorithms.
        In *Proceedings of Conference on Very Large Scale Integration: Architecture, Design, Fabrication*, pages
            509-525. California Institute of Technology, January, 1979.

[18]   Hong, J.-W. and Kung, H.T.
        I/O Complexity: The Red-Blue Pebble Game.
        In *Proceedings of the Thirteenth Annual ACM Symposium on Theory of Computing*, pages 326-333.
            ACM SIGACT, May, 1981.

[19]   Huffman, D.A.
        The Synthesis of Linear Sequential Coding Networks.
        In Cherry, C. (editor), *Information Theory*, pages 77-95. Academic press, 1957.

[20]   Kung, H.T.
       Let's Design Algorithms for VLSI Systems.
       In *Proceedings of Conference on Very Large Scale Integration: Architecture, Design, Fabrication*, pages
          65-90. California Institute of Technology, January, 1979.
       Also available as a CMU Computer Science Department technical report, September 1979.

[21]   Kung, H.T.
       Special-Purpose Devices for Signal and Image Processing: An Opportunity in VLSI.
       In *Proceedings of the SPIE, Vol. 241, Real-Time Signal Processing III*, pages 76-84. The Society of
          Photo-Optical Instrumentation Engineers, July, 1980.

[22]   Kung, H.T.,Ruane, L.M., and Yen, D.W.L.
       A Two-Level Pipelined Systolic Array for Convolutions.
       In Kung, H.T., Sproull, R.F., and Steele, G.L., Jr. (editors), *VLSI Systems and Computations*, pages
          255-264. Computer Science Department, Carnegie-Mellon University, Computer Science Press,
          Inc., October, 1981.

[23]   Kung, H.T.
       Use of VLSI in Algebraic Computation: Some Suggestions.
       In Wang, P.S. (editor), *Proceedings of the 1981 ACM Symposium on Symbolic and Algebraic
          Computation*, pages 218-222. ACM SIGSAM, August, 1981.

[24]   Kung, H.T.
       Why Systolic Architectures?
       Computer Magazine 15(1):37-46, January, 1982.

[25]   Kung, H.T. and Lehman, P.L.
       Systolic (VLSI) Arrays for Relational Database Operations.
       In *Proceedings of ACM-SIGMOD 1980 International Conference on Management of Data*, pages
          105-116. ACM, May, 1980.
       Also available as a CMU Computer Science Department technical report, August 1979.

[26]   Kung, H.T. and Leiserson, C.E.
       Systolic Arrays (for VLSI).
       In Duff, I. S. and Stewart, G. W. (editors), *Sparse Matrix Proceedings 1978*, pages 256-282. Society for
          Industrial and Applied Mathematics, 1979.
       A slightly different version appears in *Introduction to VLSI Systems* by C. A. Mead and L. A. Conway,
          Addison-Wesley, 1980, Section 8.3.

[27]   Kung, H.T. and Picard, R.L.
       Hardware Pipelines for Multi-Dimensional Convolution and Resampling.
       In *Proceedings of the 1981 IEEE Computer Society Workshop on Computer Architecture for Pattern
          Analysis and Image Database Management*, pages 273-278. IEEE Computer Society Press, Novem-
          ber, 1981.

[28]   Kung, H.T. and Song, S.W.
       A Systolic 2-D Convolution Chip.
       In Preston, K., Jr. and Uhr, L. (editor), *Multicomputers and Image Processing: Algorithms and
          Programs*, pages 373-384. 1982.
       An extended abstract appears in *Proceedings of 1981 IEEE Computer Society Workshop on Computer
          Architecture for Pattern Analysis and Image Database Management*, November 11-13, 1981, pp.
          159-160.

[29]   Lehman, P.L.
       A Systolic (VLSI) Array for Processing Simple Relational Queries.
       In Kung, H.T., Sproull, R.F., and Steele, G.L., Jr. (editors), *VLSI Systems and Computations*, pages
          285-295. Computer Science Department, Carnegie-Mellon University, Computer Science Press,
          Inc., October, 1981.

[30]   Leiserson, C.E.
       Systolic Priority Queues.
       In *Proceedings of Conference on Very Large Scale Integration: Architecture, Design, Fabrication*, pages
          199-214. California Institute of Technology, January, 1979.
       Also available as a CMU Computer Science Department technical report, April 1979.

[31]   Leiserson, C.E. and Saxe, J.B.
       Optimizing Synchronous Systems.
       In *Proceedings of the 22nd Annual Symposium on Foundations of Computer Science*, pages 23-36.
          IEEE Computer Society, October, 1981.

[32]   Liu, K.Y.
       Architecture for VLSI Design of Reed-Solomon Encoders.
       In *Proceedings of the Second Caltech VLSI Conference*. Caltech, January, 1981.

[33]   Lyon, R.F.
       Two's Complement Pipeline Multipliers.
       IEEE Transactions on Communications COM-24(4):418-425, April, 1976.

[34]   Mead, C.A. and Conway, L.A.
       Introduction to VLSI Systems.
       Addison-Wesley, Reading, Massachusetts, 1980.

[35]   Mead, C.A., Pashley, R.D., Britton, L D., Daimon, Y.T., and Sando, S.F.
       128-Bit Multicomparator.
       IEEE Journal of Solid-State Circuits SC-11(5):692-695, October, 1976.

[36]   Mukhopadhyay, A.
       Hardware Algorithms for Nonnumeric Computation.
       IEEE Transactions on Computers C-28(6):384-394, June, 1979.

[37]   Noyce, R.N.
       Hardware Prospects and Limitations.
       In Dertouzos, M.L. and Moses, J. (editor), *The Computer Age: A Twenty-Year View*, pages 321-337.
          IEEE, 1979.

[38]   Ottmann, T., Rosenberg, A.L. and Stockmeyer, L.J.
       A Dictionary Machine for VLSI.
       Technical Report RC 9060 (#39615), IBM Thomas J. Watson Research Center, Yorktown Heights,
          New York, 1981.

[39]   Peterson, W.W. and Weldon, E.J., Jr.
       Error-Correcting Codes.
       MIT Press, Cambridge, Massachusetts, 1972.

[40] Savage, C.
A Systolic Data Structure Chip for Connectivity Problems.
In Kung, H.T., Sproull, R.F., and Steele, G.L., Jr. (editors), *VLSI Systems and Computations*, pages 296-300. Computer Science Department, Carnegie-Mellon University, Computer Science Press, Inc., October, 1981.

[41] Schirm IV, L.
Multiplier-Accumulator Application Notes.

[42] Song, S.W.
On a High-Performance VLSI Solution to Database Problems.
PhD thesis, Carnegie-Mellon University, Computer Science Department, July, 1981.
Also available as a CMU Computer Science Department technical report, August 1981.

[43] Sutherland, I.E. and Mead, C.A.
Microelectronics and Computer Science.
Scientific American 237(3):210-228, September, 1977.

[44] Swartzlander, E.E., Jr. and Gilbert, B.K.
Arithmetic for Ultra-High-Speed Tomography.
IEEE Transactions on Computers C-29(5):341-354, May, 1980.

[45] Symanski, J.J.
Progress on a Systolic Processor Implementation.
In *Proceedings of SPIE Symposium, Vol. 341, Real-Time Signal Processing V*. The Society of Photo-Optical Instrumentation, May, 1982.

[46] Todd, S.
Algorithm and Hardware for a Merge Sort Using Multiple Processors.
IBM Journal of Research and Development 22(5):509-517, September, 1978.

[47] Weiser, U. and Davis, A.
A Wavefront Notation Tool for VLSI Array Design.
In Kung, H.T., Sproull, R.F., and Steele, G.L., Jr. (editors), *VLSI Systems and Computations*, pages 226-234. Computer Science Department, Carnegie-Mellon University, Computer Science Press, Inc., October, 1981.

[48] Whiteside, R.A., Hibbard, P.G. and Ostlund, N.S.
Systolic Algorithms for Monte Carlo Simulations.
Draft, CMU Computer Science Department.

[49] Yen, D.W.L. and Kulkarni, A.V.
The ESL Systolic Processor for Signal and Image Processing.
In *Proceedings of the 1981 IEEE Computer Society Workshop on Computer Architecture for Pattern Analysis and Image Database Management*, pages 265-272. November, 1981.

Hardware Algorithms for VLSI Systems

Hiroto YASUURA and Shuzo YAJIMA

Department of Information Science
Faculty of Engineering
Kyoto University
Kyoto 606, JAPAN

1. INTRODUCTION

Recent rapid advances of technology of integrated circuits make it possible to implement a logic circuit consisting of hundreds of thousands of components. Many large and complicate systems will be implemented using VLSI chips and they will seriously influence industries and various area of our daily life.

In order to design large, complicate and efficient system successfully, it is one of the pressing problems in computer science to establish a new logic design technology as well as device technology for a large and complicate hardware. Logic design will be drastically changed for large systems implemented by VLSI, since the design of large hardware containing more than hundreds of thousands of logic elements should be essentially different from existing designs in quality[1]-[3].

In the new logic design technology, several problems should be resolved. The following points are pointed out by some researchers as major problems and many researches have been pursued in industries and universities[1]-[4].

(1) High-performance: The new logic design technology should provide a logic designer with a method to design efficient highly parallel systems successfully. Logic design will be done in hierarchical manner similar to the structured design for large software. Design and analysis techniques of hardware algorithms will also be very important both practically and theoretically as

well as software. Fast algorithms, small area algorithms and low power consumption ones will be developed for various problems[1]-[8].
(2) Design effort and cost: Logic design must be effectively supported by design tools for reducing design efforts and cost. Many techniques for design automation or computer aided logic design have been developed and used practically in logic design, design verification, layout and so on. A lot of algorithms for design automation have been studied and several special purpose hardwares for simulation and routing have been proposed[9][10]. Another method for reducing design effort and cost is uniform design techniques using PLA's (Programmable Logic Arrays) and gate arrays. Researches on design of compact PLA's and optimum gate arrays have been carried out[11]. Several results of these researches are utilized in practical LSI or VLSI design.
(3) Reliability: In a large and complicate system, it is inherently difficult to test faults in the system. Some new design methods for testability, called built-in-testing and autonomous testing, have been proposed, in which circuits are designed as easily testable by embedding some mechanisms for testing in itself[12][13]. Fault tolerancy or fault recovery techniques are also important in the new logic design. Logic design also ought to be easily verifiable for design errors. Easily verifiable design methods and new verification techniques will be established.

In this article, we are concerned with design and analysis of hardware algorithms for VLSI systems. Design of good hardware algorithms is the key problem of logic design of highly parallel efficient VLSI systems. Various hardware algorithms have been developed for practically important problems such as sorting, arithmetic operations, matrix arithmetics, fast Fourier transformation and pattern matching[1]-[7]. Kung and his co-workers advocate a class of hardware algorithms suitable for VLSI implementation, named systolic algorithms[6][7]. They designed dozens of algorithms for practical problems in various fields. We also proposed a design methodology for VLSI hardware algorithms, called the Bus Connected Cellular Array algorithms (abbreviated BCA), detail of which is discussed in Chapter 4.

Design tools and analysis techniques for hardware algorithms are required for successful design. Theoretical aspects on the evaluation and analysis of efficiency of hardware algorithms have been actively discussed for these several years. Several models for parallel computation are used for these discussion, such as combinational logic circuits, VLSI models and so on[14]-[16]. On these model, various kinds of complexity measures, the number of computation or memory elements, time delay of the computation, the number of connection lines and the area of VLSI chips, were proposed and the relations between them were examined. Moreover, using these measures, many results on the complexity of various practical problems were obtained as well as the complexity theory of software. This complexity theory of hardware algorithms are discussed in the next chapter.

Regarding design tools of hardware algorithms, there are some points to be resolved. A standard model of parallel computation on which hardware algorithms are designed and analyzed should be defined. A hardware algorithm description language and verification techniques are also primary examples of the points.

In this paper, our discussion is focused on the design of high-speed algorithms by combinational circuits and hardware algorithms based on BCA. In the next chapter, an overview of the complexity theory of hardware algorithms on the combinational circuits and a VLSI model is described. In chapter 3, a few examples of high-speed circuits for arithmetic operations are presented. BCA is discussed in Chapter 4 with some considerations on the conditions of hardware algorithms suitable for VLSI implementation.

2. THE COMPLEXITY THEORY OF HARDWARE ALGORITHMS

2.1 Models and Complexity Measures

In the theory of software algorithms, a Turing machine and a Random Access Machine (RAM) are often used as models of actual computers. Time complexity and space complexity of algorithms are defined on these models and the asymptotic behaviors of these complexities are discussed[17].

For hardware algorithms, several models of computation have been proposed, such as a combinational circuit, VLSI models and multi-processor models. In this chapter, we are concerned with a combinational circuit, which is the simplest and most fundamental model of parallel computation, and a VLSI model. A number of VLSI models were proposed by Thompson[15], Brent-Kung[16], Chazelle-Monior[18], Kedem-Zorat[19], Vuillemin[20], etc. Our VLSI model in this paper is a modification of Brent-Kung's model.

In combinational circuit, the number of gates and the delay time are the most important complexity measures. These measures were examined in detail by many researchers and the relations between these measures and time and space of a Turing machine have been discussed[14]. The area is a new measure introduced with VLSI models. Other complexity measures have been also discussed and the relations between these measures began to be illuminated.

2.2 The Complexity on Combinational Circuits

Combinational circuits are one of the most fundamental circuits in digital systems. A combinational circuit computes a function $f:\{0,1\}^n->\{0,1\}^m$, which is a set of logic functions $\{f_i \mid f_i:\{0,1\}^n->\{0,1\}$ for $i=1,2,...,m\}$. The combinational circuit can be regarded as a model of parallel computation of the function f.

A finite set of logic functions $B = \{f_i \mid f_i:\{0,1\}^{n_i} ->\{0,1\}\}$ is called a basis. We only consider a complete basis. A combinational circuit C over the basis B is a labeled acyclic directed graph, each node having indegree 0, 1 or n_i, where $i=1,2,...,|B|$. Nodes with indegree 0 are input nodes each of which is labeled with an input variable x_i or a constant 0 or 1. Nodes with outdegree 0 is called a output nodes. The indegree of any output node is 1. Other nodes are called computation nodes. Each computation node with indegree n_i is uniquely labeled with an f_i in B. We can decide the output function of each computation node in the combinational circuit for a labeling of input nodes. We define a label of each output node N by the output function of the computation node which is adjacent to N. We say the combinational circuit C computes a function $f:\{0,1\}^n->\{0,1\}^m$, if there exist a labeling of input nodes and a set of output nodes $\{N_1, N_2,..., N_m\}$ such that the label of N_i equals to f_i in f, for $i=1,2,...,m$.

The size of a combinational circuit C, denoted size(C), is the number of computation nodes in C. The depth of C, denoted depth(C), is the maximum number of computation nodes on a path from an input node to an output node. When we assume that the delay of computation is only decided by the delay in computation node independent of the connections, depth(C) is a measure of the computation time of C.

The combinational complexity of a function f relative to a basis B, denoted $C_B(f)$, is the smallest size of a circuit over B that computes f. The delay complexity of a function f relative to B, denoted $D_B(f)$, is the smallest depth of a circuit over B that computes f. It is known that for almost all n-variable logic function f,

$$C_B(f) = \theta(2^n/n),$$

and

$$D_B(f) = \theta(n) \qquad [14].$$

From the view point of logic design, we are interested in the class of logic functions which can be computed by combinational circuits with depth O(log n) and polynomial size. Symmetric functions and threshold functions are contained in this class[21].

Many practical functions, such as the logical sum, the logical product, the parity function, the integer addition and the integer multiplication, can be defined for an arbitrary input length. Each of these practical functions is specified as a set of functions according to a rule independent of the number of input variables. Using this rule, we can design an effective hardware algorithm for each function. In order to deal with such a set of functions generally, we introduced a logic function sequence which is a set of logic functions containing exactly one n-variable logic function for every positive integer n[21].

[Definition 1] A logic function sequence, denoted $\{f_n\}$, is an infinite set of logic functions which contains exactly one n-variable logic function f_n for every positive integer n.

A logic function sequence is an infinite set. It is important to supply a useful representation method of function sequences. For this purpose, we utilize the formal language theory which supplies a powerful representation method of infinite sets.

[Definition 2] For a logic function sequence $\{f_n\}$, a generating language of $\{f_n\}$ is defined by a set of strings over $\{0,1\}$

consisting of $x_1 x_2 \ldots x_n$ where the vector $(x_1, x_2, \ldots, x_n)$ is in the on-set of f_n in $\{f_n\}$.

By this definition, we can easily show that there is a one-to-one correspondence between logic function sequences and formal languages over $\{0,1\}$. For example, the generating language of the function sequence $\{x_1 \oplus x_2 \oplus \ldots x_n \mid n=1,2,\ldots\}$ is the regular set represented by the regular expression $0*1*(0*10*1)*0*$.

When we examine properties of logic function sequences, we can use results of the theory of formal languages and automata. We can also discuss the relation between time and space of Turing machines and size and depth of combinational circuits[21][22]. The basic technique is the simulation of the behavior of an automaton recognizing a language L by combinational circuits which compute logic function sequence generated from L.

Unger showed that a combinational, unirateral, one-dimensional, iterative circuit (CUODIC) can be reconstructed to a combinational, iterative tree (CIT)[23]. The CUODIC is regarded as an expansion of a sequential circuit or a finite automaton. By this reconstruction, we can obtain a CIT with depth $O(\log n)$ which simulates the behavior of the sequential circuit or the automaton for an input of length n. This technique can be generally applied to function sequences generated from regular sets.

[Theorem 1] For a logic function sequence $\{f_n\}$ generated from a regular set over $\{0,1\}$, one can construct a combinational circuit with depth $O(\log n)$ and size $O(n)$ which computes f_n in $\{f_n\}$ for all n.

The construction method is as follows:
(1) Let R be a regular set over $\{0,1\}$ and $\{f_n\}$ be a function sequence generated from R. Consider a finite automaton M = $(\{0,1\}, Q, \delta, q_0, F)$ which recognizes R, where $\{0,1\}$ is the input alphabet, Q is a finite set of states, δ is the transition function mapping $Q \times \{0,1\}$ to Q, q_0 in Q is the initial state, and $F \subsetneq Q$ is the set of final states.
(2) Consider a monoid consisting of all mappings from Q to Q itself. We define two mappings μ_0 and μ_1 by

$\mu_0(q_i) = \delta(0, q_i)$ and $\mu_1(q_i) = \delta(1, q_i)$. Let G be a semigroup generated from $\{\mu_0, \mu_1\}$ in the monoid. G is uniquely defined from the finite automaton M. In G, the product of mappings are given as $\mu_i * \mu_j(q_k) = \mu_j(\mu_i(q_k))$.

(3) We encode all 's in G into binary codes, and design the following circuits.

(a) I circuit: A circuit computing a product $\mu_i * \mu_j$ from two mappings μ_i and μ_j.

(b) Ip circuit: A circuit generating μ_0 and μ_1 from inputs 0 and 1, respectively.

(c) Z circuit: A circuit generating the output of M. Namely, this circuit outputs 0 if $\mu(q_0) \notin F$, and 1 if $\mu(q_1) \in F$, for a given mapping μ.

These three circuits can be designed from M not depending on the length of an input strings, n.

(4) A circuit C_n computing f_n is constructed as a combinational iterative tree. The depth of C_n is proportional to the logarithms of n, because the circuit has a binary tree structure. The numbers of subcircuits I, Ip and Z are, respectively, n-1, n and 1.

The class of regular sets is included in the simplest class from the viewpoint of the complexity of logic functions. The class of regular sets contains several practically important logic function sequences such as adders, comparators, parity generators and modulo counters. The design procedure is very useful for design of high-speed circuits of these functions. For instance, several high-speed adders such as a carry look-ahead adder can be designed successfully by this method.

The most general acceptor of formal languages is a Turing machine. It is known that any function sequence generated from a language accepted by a T(n)-time bounded, S(n)-space bounded deterministic Turing machine is computed by a circuit with size O(T(n) log S(n))[14]. It is also known that an S(n)-space bounded nondeterministic Turing machine can be simulated by a circuit with depth $O(S(n)^2)$[22]. Moreover we have the following result[24].

[Theorem 2] A function sequence generated from a language accepted by a T(n)-time bounded, S(n)-space bounded deterministic Turing machine has the delay complexity O(S(n) log T(n)).

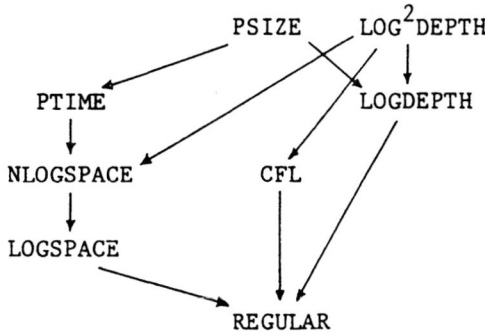

Fig.1 Relation among classes of the Turing
Machine and Circuit Complexity.

From this theorem, we say that a function sequence generated from a language accepted by a polynomial-time bounded, S(n)-space bounded Turing machine has the delay complexity O(S(n) log n).

In Fig.1, the relations among classes of the Turing machine complexity and the circuit complexity are sumarized.

2.3 The Area Complexity on a VLSI Model

In VLSI circuits, it is known that the chip area of a circuit depends on not only the number of logic elements but also the area for the wiring and input/output terminals. A number of mathematical models of VLSI are proposed and several techniques to evaluate the area are provided[15]-[20].

In this section, we introduce a boundary condition of input/output terminals and discuss methods to eveluate the area required for layout of tree circuits[25]. Our VLSI model is an extention of the model of Brent-Kung[16]. The boundary condition introduced here is that all inputs and outputs should be performed on the boundary of the convex region in which a circuit is embedded. This condition is a reasonable and realistic restriction on VLSI layout because
(1) The inputs and outputs of a VLSI chip is ordinarily performed on input/output pads located at the boundary of the chip.

(2) The connections between subcircuits in a VLSI chip is realized by wires which connects boundary of the regions of the subcircuits.

It is known that the area required for layout of a certain class of circuits is essentially different whether this condition is considered or not.

A VLSI model considered in this section is as follows:

(1) A circuit is embedded into a convex planer region R.

(2) Wires have minimum width λ (a positive constant).

(3) At most ν ($\nu \geq 2$, a constant) wires can overlap at any point of R.

(4) Each logic element contains a $\lambda \times \lambda$ square and their shapes and area are given for each sort of elements.

(5) No logic element overlaps other logic elements and wires in R.

(6) Inputs and outputs of the circuit is performed through wires on the boundary of R.

A tree circuit with degree s ($s \geq 3$) is a circuit represented by a graph G satisfying the following conditions.

(1) G is connected and does not contain a cycle (i.e. G is a tree).

(2) The degree of any input or output node in G is exactly 1.

(3) The degree of any computation node in G is not greater than s and not less than 3.

In this section, we will consider tree circuits with degree 3. If we delete an input or output node N from a tree circuit C, C becomes a binary tree whose root is the computation node just adjacent to N.

For a binary tree T, we define the effective height h of T by the following manner.

(1) The label of each leaf is 0.

(2) The label of a computation node N, denoted label(N), is calculated from the labels of its son nodes N_1 and N_2 by the following procedure:

    if label(N_1) = label(N_2)

       then label(N) := label(N_1) + 1;

       else label(N) := max{label(N_1), label(N_2)};

(3) The effective height of T is the label of the root node.

The effective height of any binary tree obtained from a tree circuit C is different each other at least one. Thus we define the effective height of the circuit C by the largest one of all binary trees obtained from C. This effective height is a measure of area required for embedding of the circuit.

[Theorem 3] The area of a convex region necessary and sufficient for embedding of a tree circuit with degree 3, which has n terminal nodes and effective height h, is θ(nh)[25].

It is easy to show a layout of the tree circuit in the region of area O(nh). This results can be generalized to a tree circuit with degree s greater than 3[25].

Next we consider the relation between depth and area of combinational circuits. We first consider a fanout-free circuit with one output node · over the basis B_2 consisting of all 2-variable functions. Let L(d,h) be the maximum number of input nodes in fanout-free combinational circuits with depth d and effective height h. On L(d,h), we have

$$L(d,h) = \sum_{i=0}^{h} {}_dC_i .$$

Using this property, we obtain

$$L(d,h) < (\tfrac{d}{2})^h, \quad (h \geq 6)$$
and
$$L(d,h) \leq 2^d - 2^{d-h-1} + \tfrac{1}{2}.$$

Then from Theorem 3 and these results, we have lower bounds of the area for a layout of a combinational circuit.

[Theorem 4] For area A required to embedding any fanout-free circuit with n input nodes, 1 output and depth d over the basis B_2, the following inequalities hold:

$$A \geq \frac{cn \log_2 n}{\log_2(d/2)} ,$$
and
$$A \geq cn \log_2(\frac{2^{d-1}}{2^d - n + (1/2)}).$$

The depth of a combinational circuit computing an n-variable function is $\Omega(\log n)$. By Theorem 4, if d=O(log n), then A=

Ω(n log n) and if d=O(n), then A= Ω(n). This shows a trade-off relation between area and time in computation by combinational circuits.

We can generalize this result to combinational circuits with fanout, since any combinational circuit over B_2 computing an n-variable function includes a binary trees with n leaves.

[Corollary 1] For any combinational circuit over B_2 computing a logic function exactly n variables, there is a trade-off relation between depth d and area A for a layout such that
$$A \log_2 d = \Omega (n \log n).$$

3. HARDWARE ALGORITHMS FOR ARITHMETIC OPERATIONS

Arithmetic operations are the most fundamental and significant operations in digital systems. Many researches on effective algorithms for arithmetic operations have been carried out and the results of these researches are utilized in implementation of practical arithmetic circuits[26].

In 1950's and 1960's, much effort was spent to reduce the number of computation elements, because hardware was too expensive and it was difficult to implement a large scale circuit. In 1970's as LSI technology advances, the restrictions on hardware realization was relaxed. Requirements of more efficient systems push many researchers to studies on high-speed circuits. A lot of high-speed arithmetic circuits using LSI and VLSI technology have been proposed.

In design of high-speed arithmetic circuits, it is important to develop good algorithms for efficient highly parallel computation. The theory of circuit complexity discussed before is available in practical design of hardware algorithms for high-speed arithmetic operations. Many theoretical results on the complexity of arithmetic operations measured by size, depth and area have been reported[1][4][16][26].

In this chapter, we briefly survey hardware algorithms for integer multiplication, integer division and square rooting.

3.1 Integer Multiplication

Integer multiplication is widely used as a basic operation in general purpose computers, in process controllers and signal processors. Many high-speed algorithms for integer multiplication in software and hardware have been proposed and used practically. In table 1, several hardware algorithms for integer multiplication are compared.

In applications not required so much high speed computation, add-and-shift multiplication is generally used. The speed of computation is improved, when a carry look-ahead adder is adopted[26]. Serial multiplication is implemented in signal processing in which operands are input serially[27]. These above algorithms are implemented by sequential circuits.

For high-speed multiplication by combinational circuits, array multiplication and matrix generation-reduction scheme are developed and implemented for practical use. Array multiplication is attractive for their compactness and regularity of its iterative array structure using one basic circuit type, but their speed of operation increases linearly with the operand length and thus slow for large words[26]. Matrix generation-reduction scheme is much faster for large operands since their speed of operation increases with the logarithm of the operand length[28]. The basic idea of this algorithm was proposed by Karatsuba and Ofman, and the most popular circuit based on it is known as Wallace's tree[29][30]. Several papers discussed about multiplier based on this algorithm with higher performance[26][28].

Recently, we developed a new hardware algorithm for integer multiplication, which uses internally a redundant binary notation[31]. In the algorithm, partial products in the multiplication are represented by a redundant binary notation using 0, 1 and -1 for each digit. Utilizing a redundancy in the notation, addition of two redundant binary numbers can be computed in constant time not depending on the length of the data. Thus we can obtain the product of the multiplication represented by the redundant notation in time proportional to the logarithm of the length of the multiplier and the multiplicand, when we construct a binary tree of adders.

Table 2 shows an evaluation of the size (the number of

Algorithm	Size	Area	Speed of Computation
Add-and-Shift Multiplication (Ripple Carry Adder)	n	n	n^2
Add-and-Shift Multiplication (Carry Lookahead Adder)	n	$n \log n$	$n \log n$
Serial Multiplication	n	n	n
Array Multiplication	n^2	n^2	n
Matrix Generation-Reduction Multiplication	n^2	$n^2 \log n$	$\log n$
Redundant Binary Addition Algorithm	n^2	$n^2 \log n$	$\log n$
Brent-Kung's Algorithm	–	$n \log n$	$\sqrt{n} \log n$
Shonhage-Strassen's Algorithm	$n \log n \log \log n$	–	$\log n$

n:the length of operand

Table 1 Hardware Algorithms for Integer Multiplication

n	8	16	32	64
Array Algorithm	29/528	61/2336	125/9792	253/40064
Matrix Generation- Reduction	22/672	24/2516	30/9064	34/35207
Redundant Binary Addition	22/623	28/2535	33/9948	39/38842

(depth/size) 4-input NOR/OR gates

Table 2 Depth and Size of Multipliers

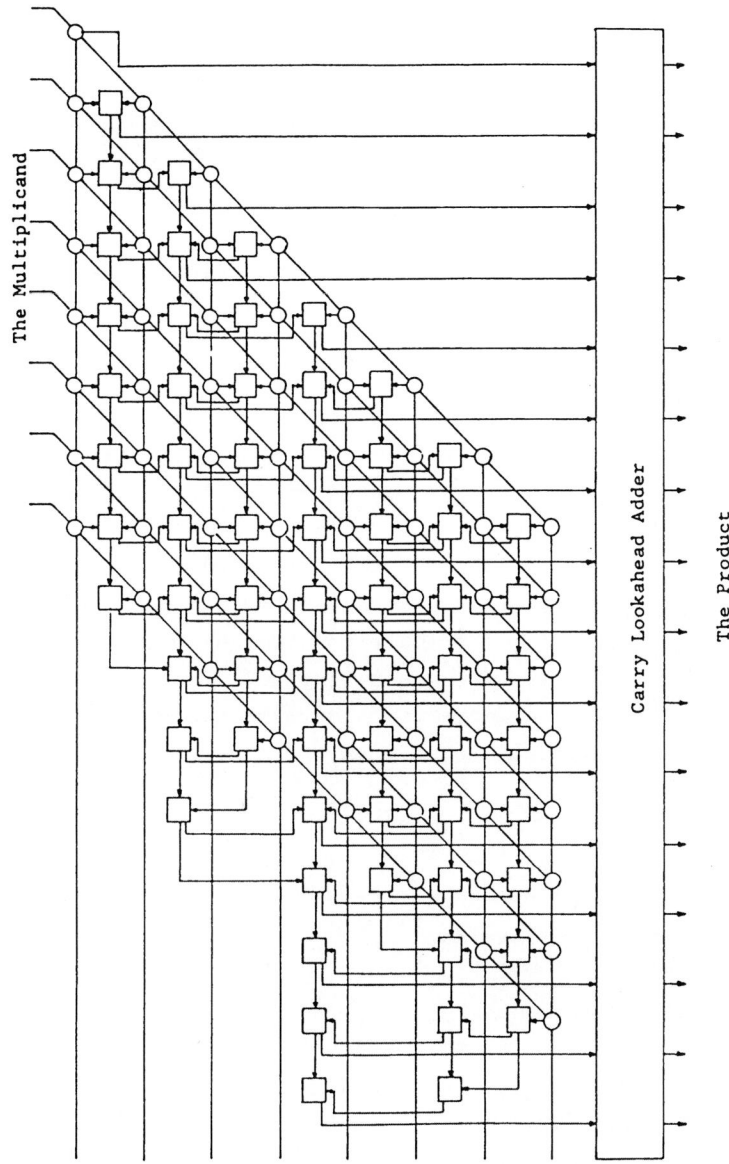

The Multiplicand

The Multiplier

Carry Lookahead Adder

The Product

Fig.2 An Array Multiplier Using Redundant Binary Notation.

gates) and the depth (the speed of computation) of circuits designed according to these three algorithms. Gates used in the design are 4-input NOR/OR gates. The circuit based on matrix generation-reduction algorithm is realized as a combination of Booth's algorithm, Wallece's tree and a carry look-ahead adder. The depth of matrix generation-reduction and redundant binary addition algorithms are extremely smaller than the array multiplier for large n. This example shows that the design of a good hardware algorithm results in tremendous improvement on efficiency.

Although the size of circuits of these three algorithms are the same order, namely $O(n^2)$, the upper bounds of the area on VLSI have quite different order, $O(n^2)$ for array multiplication and $O(n^2 \log n)$ for matrix generation-reduction and redundant binary addition ones. This difference is caused by the difference of the complexity of interconnection in the circuits. Layouts of circuit based on matrix generation-reduction scheme are complicated and it will be difficult to find the optimum layout. On the other hand, a simple and regular layout of a circuit based on redundant binary addition algorithm can be easily found (See Fig.2). This algorithm has both two majors, the regularity and high-speed.

Brent-Kung's algorithm in table 1 achieves the best upper bound of the area-time product[16]. Shonhage-Strassen's one is the best upper bound of the number of Boolean operations required to n-bit integer multiplication[17].

3.2 Division and Square Rooting

In this section, we are concerned with several arithmetic operations on n-bit integers including division and square rooting. We only consider the delay complexity of these operations and combinational circuits with small depth for them[32].

For any positive real number r and any positive integer m, $\lfloor r \rfloor_m$ denotes the number represented by the upper m bits of the binary representation of r.

We consider the following arithmetic operations. Assume that P is an n-bit integer.
(1) Division: Dividing a 2n-bit integer by an n-bit integer and

producing an integer part of the quotient.

(2) Inversion: Computing $\lfloor 1/P \rfloor_n$.

(3) Square rooting: Computing $\lfloor \sqrt{P} \rfloor_n$.

(4) Inversion of square root: Computing $\lfloor 1/\sqrt{P} \rfloor_n$.

(5) k-th rooting: For a positive integer k, computing $\lfloor \sqrt[k]{P} \rfloor_n$.

(6) Inversion of k-th root: For a positive integer k, computing $\lfloor 1/\sqrt[k]{P} \rfloor_n$.

(7) n-th power: Computing $\lfloor P^n \rfloor_n$.

[Theorem 5] These seven arithmetic operations can be computed by circuits with the depth proportional to $(\log n)^2$.

Proof: The algorithms for (2), (4) and (6) are an application of Newton's iteration, i.e.,

$$a_2 = \lfloor \tfrac{1}{k}((k+1)a_1 + a_1^{k+1}P) \rfloor_n,$$

where k=1 for (2) and k=2 for (4). Suppose P_1 is the integer represented by upper n/2 bits of P. If we adopt $\lfloor 1/\sqrt[k]{P_1 2^{n/2}} \rfloor_n$ as the initial approximation of a_1, we can show that the error between a_2 and $\lfloor 1/\sqrt[k]{P} \rfloor_n$ is independent of n. The circuit based on the algorithm have the depth proportional to $(\log n)^2$. Since $P(1/\sqrt[k]{P})^{k-1} = \sqrt[k]{P}$, we can compute (3) and (5) from the results of (4) and (6), respectively, by circuits of depth $O((\log n)^2)$. (1) can be clearly computed from the result of (2) as $x/y = x(1/y)$. Since squaring has the delay complexity of $O(\log n)$ and one can compute $\lfloor P^2 \rfloor_n$, $\lfloor P^4 \rfloor_n$, $\lfloor P^8 \rfloor_n$, ..., $\lfloor P^n \rfloor_n$ by at most $\lceil \log n \rceil$ times of square rooting, (7) is also computed by a circuit with depth $O((\log n)^2)$.   Q.E.D.

For the combinational complexity, we know that square rooting, inversion and division are both as complex as multiplication[33]. The best known upper bound on the combinational complexity of these operations is $O(n \log n \log\log n)$ [17]. Moreover, we can easily extend Alt's method[33] to prove that k-th rooting has the same complexity. Therefore the above six operations except the n-th power have the same combinational complexity, i.e., if there exists an algorithm with size S(n) for one of these operations, or for multiplication, one can construct algorithms for the others whose size are also $O(S(n))$. Note that the size of circuits for (1)-(6) considered in the proof of

Theorem 5 are both O(n log n loglog n) when we use a multiplication algorithm with size O(n log n loglog n).

A reduction scheme for these arithmetic operations on the delay complexity is shown in Fig.3. An arrow means that if the operation placed on the tail of the arrow can be computed by a circuit with depth $O(D(n))$, then the operation pointed by the arrow can be also computed in depth $O(D(n))$.

(a) is obvious. (b) is shown by $P = P(1/P)$ and (d). (c) is also shown by the manner similar to (b). (d) is shown by the following formula,

$$\lfloor \frac{2^{2n}}{P} \rfloor_n = \lfloor \sqrt{2^{4n}P + \lfloor \sqrt{2^{8n}(P+1)} \rfloor} \rfloor - \lfloor \sqrt{2^{4n}P + \lfloor \sqrt{2^{8n}P} \rfloor} \rfloor + O(2^{n/2}).$$

(e) is trivial because square rooting is a special case of k-th rooting. (f) is shown by the expansion into power series such as

$$\lfloor x^j \rfloor_n = \lfloor \sum_{i=0}^{n} \frac{j(j-1)\ldots(j-i+1)}{i!} (1-x)^i \rfloor_n,$$

where $j = -1, \pm(1/2), \pm(1/3), \ldots$ and $1/2 \leq x < 1$.

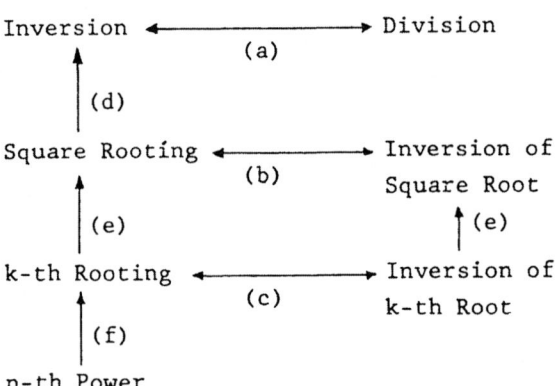

Fig.3 Reduction Scheme for Arithmetic Operations.

4 HARDWARE ALGORITHMS ON BUS CONNECTED CELLULAR ARRAYS

Hardware algorithms for VLSI circuits which process problems dealing with a large number of data are desirable to have the following properties.

(1) High-performance: The algorithm achieves high-performance using efficient pipelining and multiprocessing. Input/output and processing are effectively overlapped each other.

(2) Low cost design: Logic design and its verification should be easy. Layout of the circuit demands the significant consideration in the design of the hardware algorithm.

(3) Simple and efficient interface: The algorithm has a simple and efficient interface to the system in which the circuit is used as a subsystem.· The restriction on the number of I/O ports, the amount of communications between the system and the circuit, and the complexity of protocols of the communication invite deep consideration in the algorithm design. These factors sometimes determine the performance of the whole system.

(4) Restrictions and extensibility: Restrictions on the processing ability of the circuit caused by the size are relaxed as much as possible. Processing time is desirable to depend on only the size of input data not on the size of the circuit. The algorithm is easily extensible. To process a large number of data exceeding the ability of the circuit, an effective algorithm using the circuit more than once is provided.

We developed hardware algorithms which are realized on Bus Connected Cellular Arrays (BCA)[34]-[36]. These algorithms posess the following properties.

(1) The algorithm is implemented by a linear array structure of a few different types of simple cells.

(2) The algorithm uses pipelining and multiprocessing. Input/output and processing are completely overlapped.

(3) The communication structure of the circuit is simple and regular. Local communication between neighbour cells and global communication using one-directional buses are efficiently performed and controlled by a simple control circuit in each cell.

(4) Interface of the algorithm is very simple. Inputs and outputs are performed sequentially.

(5) Since a linear array structure is adopted, the circuit can be easily expanded and the number of pins of a VLSI chip is independent of the number of cells on the chip.

(6) The processing time is linearly proportional to the number of data.

Our algorithms on BCA is similar to the systolic algorithms proposed by Kung and his group[6][7]. The most significant difference between these two kinds of algorithms is that in BCA algorithms global communications by buses are permitted. The global communications improve the speed of algorithm drastically and reduce the complexity of communication.

Algorithms must be designed under realistic assumptions on input/output protocols. Highly parallel input and output increases infeasiblly the complexity of communication of the outside of the algorithm, though the algorithm seems to achieve high performance. In BCA, inputs and outputs are performed sequentially, and their protocol is very simple.

Hardware algorithms are inherently restricted their ability of processing by the size of circuits. However, algorithms are desired to process problems smaller than their ability in time proportional to the size of problems. On BCA algorithms, this property can be easily realized using global bus communications.

We proposed a sorting algorithm on BCA, called the parallel enumeration sort[34][35]. This algorithm can be introduced to conventional computer systems without changing their architecture. The processing time is linearly proportional to the number of data for sorting. The sorting circuit consists of a linear array of one type of simple cells each of which includes two registers, a comparator and a counter. These cells are connected by two buses (See Fig.4). Since the circuit is extensible only by connecting the same circuit, we can implement a large circuit connecting VLSI chips including the circuits.

The sorting circuit is fed a sequence of keys x_1, x_2, ..., x_n serially, and outputs a sequence of orders c_1, c_2, ..., c_n, where c_i is the order of x_i in the input sequence. Fig.5 shows a flow of the parallel enumeration sorting on the circuit. The input sequence is provided consequently from t_1 to t_n and the output sequence is returned from t_{n+1} to t_{2n}. At time t_i, x_i

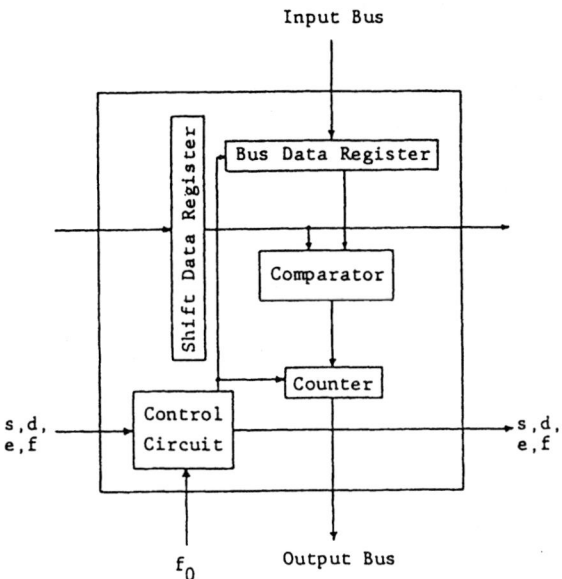

Block Diagram of the Cell.

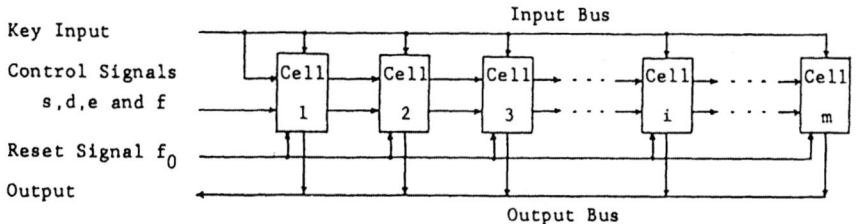

Fig.4 A Parallel Enumeration Sorting Circuit.

arrives at the input terminal and is transported to the first cell and i-th cell through the input bus. After receiving x_i, the i-th cell begins comparison and counting. At time t_{n+i}, the i-th cell completes counting and transmits the counting result c_i to the output terminal through the output bus.

We can easily perform rearrangement of records corresponding keys using the sequence of the orders obtained by this algorithm. Since the sorting circuit processes only keys, not whole records for sorting, the communication between the sorting circuit and

125

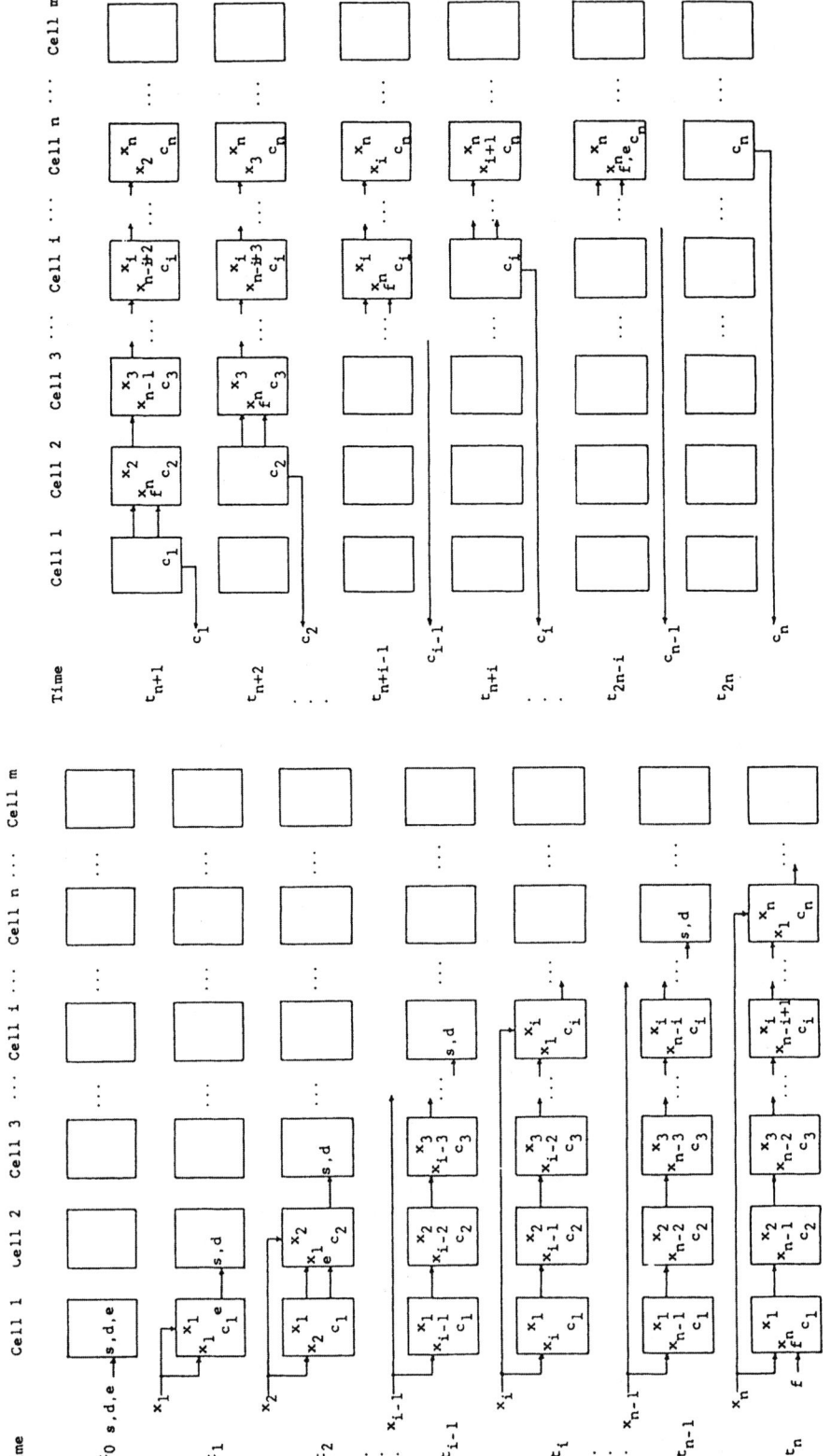

Fig.5 The Flow of Parallel Enumeration Sorting

memory devices which store the original records are minimized.

We also developed BCA algorithms for pattern matching, matrix multiplication, very long integer multiplications and join operation in relational databases[36][37].

5. CONCLUSION

In this paper, we discussed several problems in the design and analysis of hardware algorithms. The complexity theory of logic circuits and parallel computation will form the theoretical foundation of design and analysis of hardware algorithms which will become more important for large VLSI systems.

A formal description method of hardware algorithms will be proposed on a general model of parallel computation. Design tools for hardware algorithms will be developed such as a hardware description language, an automatic translation system from the language to circuits, a verification support system for algorithms, etc. Results of theoretical researches on the circuit complexity will play an important role in the development of these design tools.

Many hardware algorithms will be designed to various problems. Design methodologies such as systolic algorithms or BCA algorithms will become more important and be studied. Since a large system is a combination of software and hardware, a design methodology including design of both software and hardware algorithms will be discussed.

Acknowledgement

The authors would like to express their sincere appreciation to Mr.N.Takagi for his many helpful co-operation in preparing the manuscript. The authors would like to thank Prof.Y.Kambayashi and the member of the Yajima Lab. of Kyoto University for their discussions.

References

[1] S.Yajima and H.Yasuura, "Hardware Algorithms and Logic Deign Automation --- An Overview and Progress Report ---", Kyoto Univ. Yajima Lab. Research Rep. ER 82-01, June 1982. Also to appear in Lecture Notes on Computer Science, Springer-Verlag.

[2] "Highly Parallel Computing" Edited by L.S.Hayens, IEEE Computer, vol.15, no.1, pp.7-96, Jan. 1982.

[3] S.Yajima, H.Yasuura and Y.Kambayashi, "Design of Hardware Algorithms and Related Problems", IECE Technical Rep. AL81-86, Dec. 1981 (in Japanese).

[4] N.Tokura, "VLSI Algorithms and Area-Time Complexity", Joho-Shori vol.23, no.3, pp.176-186, March 1982 (in Japanese).

[5] C.A.Mead and L.A.Conway, "Introduction to VLSI Systems", Addison-Wesley, Reading, Mass., 1980.

[6] H.T.Kung, "The Structure of Parallel Algorithms", Advanced in Computers, vol.19, Academic Press, 1980.

[7] M.Foster and H.T.Kung, "The Design of Special-Purpose VLSI Chips", IEEE Computer, vol.13, no.1, Jan. 1980.

[8] S.Yajima and K.Inagaki, "Power Minimization Problems of Logic Network", IEEE Trans. on Comput., vol.C-23, no.2, pp.153-165, Feb. 1974.

[9] G.F.Pfister, "The Yorktown Simulation Engine: Introduction", Proc. 19th Design Automation Conf.pp.51-54, June 1982.

[10] H.G.Adshead, "Towards VLSI Complexity: The DA Algorithm Scaling Problem: Can Special DA Hardware Help?", Proc. 19th Design Automation Conf. pp.339-344, June 1982.

[11] R.A.Wood, "A High Density Programmable Logic Array Chip", IEEE Trans. on Comput. vol.C-28, no.9, pp.602-608, Sept. 1979.

[12] S.Yajima and T.Aramaki, "Autonomously Testable Programmable Logic Arrays", Proc. FTCS-11, pp.41-43, June 1981.

[13] T.Williams and K.P.Parker, "Design for Testability -- A Survey", IEEE Trans. on Comput., vol.C-31, no.1, pp.2-15, Jan. 1982.

[14] J.E.Savage, "The Complexity of Computing", Wiley-Interscience, Reading Mass., 1976.

[15] C.D.Thompson, "Area-Time Complexity for VLSI", Proc. 11th Symposium on the Theory of Computing, pp.81-88, May 1979.

[16] R.P.Brent and H.T.Kung, "The Area-Time Complexity of Binary Multiplication", JACM, vol.28, no.3, pp.521-534, July 1981.

[17] A.V.Aho, J.E.Hopcroft and J.D.Ullman, "Design and Analysis of Computer Algorithms", Addison-Wesley, Reading, Mass., 1974.

[18] B.Chazelle and L.Monier, "Towards More Realistic Models of Computation for VLSI", Proc. 11th Symposium on the Theory of Computing, pp.209-213, April 1979.

[19] Z.M.Kedem and A.Zorat, "On Relations between Inputs and Communication/Computation in VLSI", Proc. 22nd Symp. on the Foundations of Computer Science, pp.37-44, Oct.1981.

[20] J.Vuillemin, "A Combinatorial Limit to the Computing Power of VLSI Circuits", Proc. 21st Symp. on the Foundation of Computer Sience, pp.294-300, Oct. 1980.

[21] H.Yasuura, "Theory of Complexity of Logic Functions and its Application to Logical Design of High-Speed Logic Circuits", Trans. of the Information Processing Society of Japan, vol.21, no.4, pp.268-278, July 1980 (in Japanese).

[22] A.Borodin, "On Relating Time and Space to Size and depth", SIAM Jornal of Computing, vol.6, no.4, pp.733-744, Dec. 1977.

[23] S.H.Unger, "Tree Realizations of Interactive Circuits", IEEE Trans. on Comput., vol.C-26, no.4, pp.365-383, April 1977.

[24] H.Yasuura, "Width and Depth of Combinational Logic Circuits", Information Processing Letters, vol.13, no.4,4,end, pp.191-194, 1981.

[25] H.Yasuura and S.Yajima, "On the Area of Logic Circuits in VLSI", Trans. of Institute of Electronics and Communication Engineers of Japan, vol.J65-D, no.8, Aug. 1982.

[26] K.Hwang, "Computer Arithmetic:Principle, Architecture and Design", John-Wiley & Sons, Reading, Mass., 1979.

[27] L.B.Jackson, S.F.Kaiser and H.S.McDonald, "An Approach to the Implementation of Digital Filters," IEEE Trans. Audio Electro., AU-16, Sept. 1968.

[28] W.J.Stenzel, W.J.Kubitz and G.H.Garcia, "A Compact High-Speed Parallel Multiplication Scheme," IEEE Trans. on Comput., vol.C-26, no.10, pp.948-957, Oct.·1977.

[29] A.Karatsuba and Y.Ofman, "Multiplication of Multidigit Numbers with Computers", Dokl. Akad. Nauk. SSSR, no.145, Feb. 1962.

[30] C.S.Wallace, "A Suggestion for a Fast Multiplier", IEEE Trans. on Electro. Comput., vol EC-13, no.1, pp.14-17, Feb. 1964.

[31] N.Takagi, H.Yasuura and S.Yajima, "A High-Speed Array Multiplier Using Redundant Binary Representation," to appear.

[32] H.Yasuura and S.Yajima, "On the Delay Complexity of Square Rooting in Combinational Logic Circuits", Tech. Rep. of IECEJ, AL79-29, July 1979.

[33] H.Alt, "Square Rooting is as Difficult as Multiplication", Computing, vol.21, pp.221-232, Jan.1979.

[34] H.Yasuura and N.Takagi, "A High-Speed Sorting Circuit Using Parallel Enumeration Sort", Trans. IECE, vol.J65-D, no.2, pp.179-186, Feb.1982 (in Japanese).

[35] H.Yasuura, N.Takagi and S.Yajima, "The Parallel Enumeration Sorting Scheme for VLSI", to appear in IEEE Trans. on Computer vol.C-31, no.12, Dec.1982.

[36] H.Yasuura, "Hardware Algorithms for VLSI", Proc. Joint Conf. of 4 Institutes Related on Electric Engineering, 34-4, Oct. 1981 (in Japanese).

[37] H.Miyata, H.Yasuura and S.Yajima, "Hardware Algorithm for Large Integer Multiplication", Tech. Rep. of IECEJ, AL81-99, Jan.1982 (in Japanese).

Chapter 4

VLSI Design and Testing

GRAPH-BASED DESIGN SPECIFICATION OF PARALLEL COMPUTATION

Atsushi Iizawa and Tosiyasu L. Kunii

Department of Information Science,
University of Tokyo,
7-3-1 Hongo, Bunkyo-ku, Tokyo, 113 Japan

Abstract

This paper describes C-graphs that illustrate parallel computations. It also describes an Algol-like language called PCDL, which is used to code a class of parallelism we call *locally parallel processing*. The discussion also covers some problems encountered in designing *locally parallel* processors, including partitioning and serializing, which can be accomplished using PCDL.

1. Introduction

Parallel processing is the key to increasing execution speed of information processing. Communication between and among processors is a crucial point because communication overhead typically presents a bottleneck in overall execution time. To achieve a high speed computer, communication activities must be reduced. In this paper, we handle parallelism in a way that communications among processors occupyies a very small part of total computation. We call this type of parallelism *locally parallel* computation. Locally parallel computation makes it possible to produce a computer composing many processors. Although locally parallel computation is a rather restricted concept of parallelism, we find a many applications to which it can be applied. A typical example is computer graphics, where each pixel can be processed locally in parallel.

To make a model of parallel computation, we use directed graphs. Graphs are suitable for the following reasons:
   (1) Graphs can be used to represent various kinds of structures. For example, block diagrams and logic diagrams are used extensively in hardware design processes.
   (2) Graphs can be used to represent flows, such as control flows and data flows. They are especially suitable for representing parallelism.
Many models using directed graphs have been proposed for representing parallelism; typical examples are Petri nets[1] and data flow graphs[2,3]. The properties, concepts and techniques of Petri nets were developed as simple and powerful methods for describing and analyzing the flow of information and control in systems. But they are too simple to describe the behavior of practical hardware when used in their original form. Data flow graphs basically represent only data flow. This means that data flow graphs are still too abstract for hardware descriptions.

We propose C-graphs as a notation for a parallel computation model. C-graphs are defined formally with first order logic and are designed so to be easily translated into lower level descriptions using existing hardware description languages, such as ISP[4] or DDL[5].

One problem in designing locally parallel processors is the need to handle many processors, sometimes numbering in the ten thousands. Without a compact, convenient way to represent such a large number of processors, economical design is not feasible. We propose the Parallel Computation Description Language (PCDL) as a high level language for coding locally parallel computation.

To make best use of locally parallel processing in an actual system, we must be able to select any degree of parallelism. The scope of selection, however, is limited by cost performance and technological considerations. The final section of this paper discusses partitioning and serialization, which are transformations of PCDL programs.

2. Computation Graphs

We use computation graphs (C-graphs) to represent parallel computation. The model of parallel computation represented by C-graphs consists of many processors and a common memory. The following conditions are assumed to be satisfied:

1. The processors run asynchronously in parallel.
2. Any number of processors can simultaneously read the common memory.
3. Any number of processors can simultaneously write to the common memory. If several processors write to the same memory cell, however, contents cannot be guaranteed.

A *computation graph (C-graph)* is a directed graph whose components have semantics for computations executed on the above model. This section provides an informal explanation of the primitive elements of C-graphs. See Appendix A for the formal definitions. The formal definition of the C-graph's elements not only gives rigorous semantics to C-graphs, but also suggests how C-graphs can be converted to hardware descriptions in DDL.

2.1 Terminologies

This section explains the terms used to describe C-graphs.

A *directed graph* G is a triple (N,A,f), where N and A are disjoint finite sets and f is a function such that $f:A \rightarrow N \times N$. N, A and f are called *nodes, arcs* and an *arc function*, respectively. If $f(a) = (n_1,n_2)$, node n_1 is called the *source* and n_2 the *target* of the arc. Arc a is an *output* arc of node n_1 and an *input* arc of node n_2. When there is no need to distinguish between arcs and nodes of graph G, the term *components* is used to refer to the nodes or arcs. Graph $G' = (N',A',f')$ is a *subgraph* of graph $G = (N,G,f)$ if $N' \subset N$, $A' \subset A$ and $f'(a') = f(a')$ for $\forall a' \in A'$.

2.2 Informal Description of C-graphs

C-graphs represent both data flows and control flows. A distinction is made between these two types of flows. C-graphs do not specify the structure of the common memory. A specification of common memory is necessary at lower levels of description. At the C-graph level, however, the common memory is assumed to provide mapping from its addresses into values. The common memory is accessed by the two types of primitives nodes: the get node and the put node.

In C-graphs, it is assumed that all local values are retained in arcs. The arcs are divided into two classes according to the type of values they retain: *control arcs* and *data arcs*. A control arc, which retains a truth value, is set true by its source node and is set false by its target node. A control arc cannot be set true when the arc is already true. Both nodes inspect the value. For convenience, we say that a control arrives at a control arc when the arc is set true. The control is a *token* in Petri net terminology nd data flow graph terminology. Control arcs are represented by a dotted line.

The value of a data arc, which is general data such as integers or floating point

134

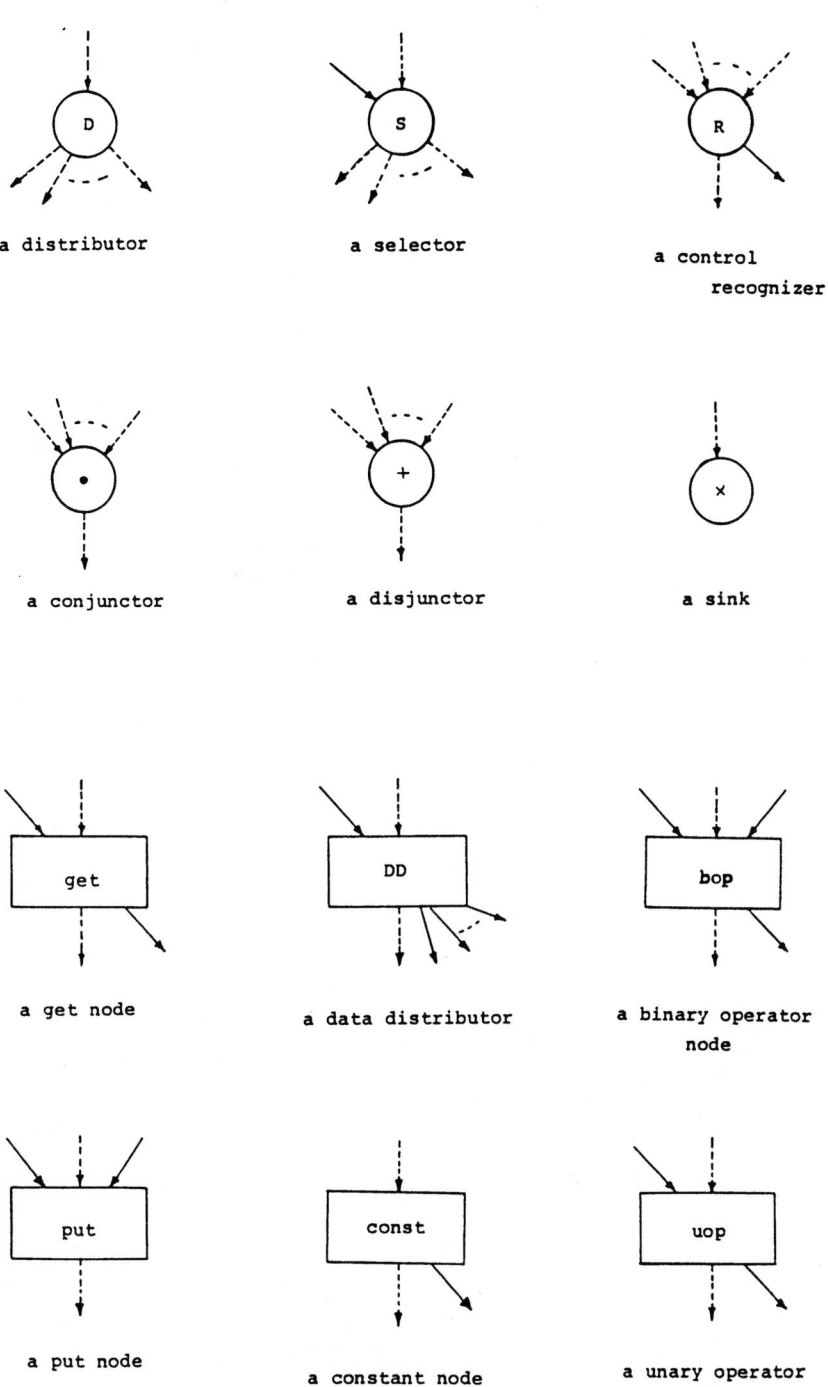

Figure 1. Graphic symbols for primitive nodes of C-graphs

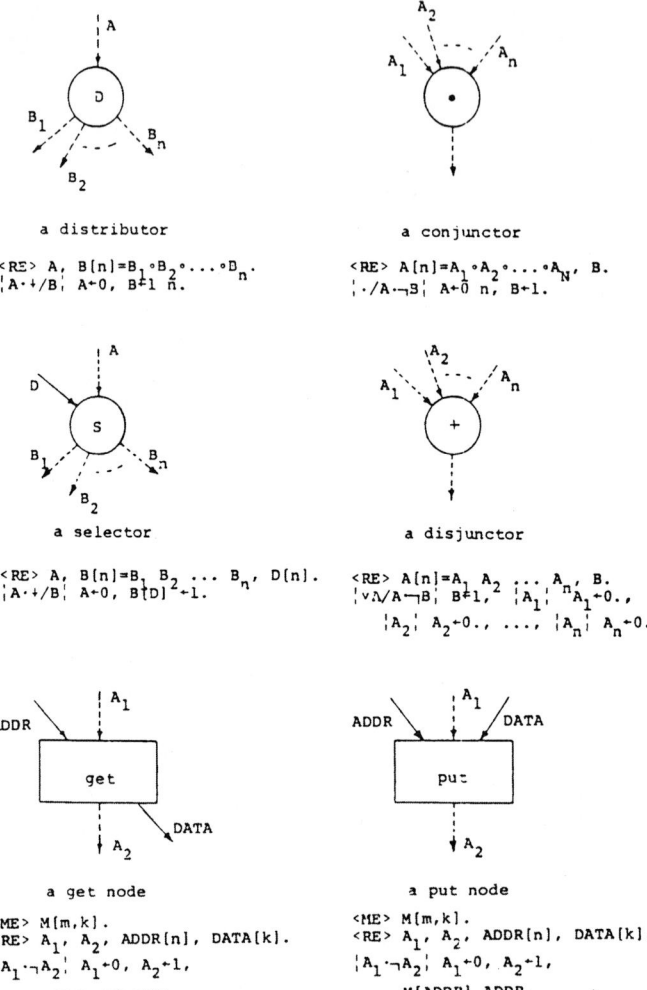

a distributor

<RE> A, B[n]=B$_1$◦B$_2$◦...◦D$_n$.
|A·+/B| A+0, B+1 ñ.

a conjunctor

<RE> A[n]=A$_1$◦A$_2$◦...◦A$_N$, B.
|·/A.¬B| A+0 n, B+1.

a selector

<RE> A, B[n]=B$_1$ B$_2$ ... B$_n$, D[n].
|A·+/B| A+0, B[D]^2+1.

a disjunctor

<RE> A[n]=A$_1$ A$_2$ ... A$_n$, B.
|v∧/A¬B| B+1, |A$_1$| A$_1$+0.,
|A$_2$| A$_2$+0., ..., |A$_n$| A$_n$+0.

a get node

<ME> M[m,k].
<RE> A$_1$, A$_2$, ADDR[n], DATA[k].
|A$_1$.¬A$_2$| A$_1$+0, A$_2$+1,

DATA M[ADDR].

a put node

<ME> M[m,k].
<RE> A$_1$, A$_2$, ADDR[n], DATA[k].
|A$_1$.¬A$_2$| A$_1$+0, A$_2$+1,

M[ADDR] ADDR.

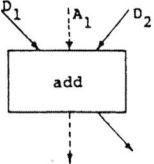

An addition node

<RE> A$_1$, A$_2$, D$_1$[n], D$_2$[n], R[n].
|A$_1$.¬ A$_2$| A$_1$ 0, A$_2$+1, R+D$_1$+D$_2$.

Figure 2. DDL representations for C-graph primitives

numbers, is assigned by the sources of the arc; the value is inspected only by the target node. Values of data arcs are assumed to be integers, floating point numbers, Boolean values and addresses of the common memory. Data arcs are represented by a solid line.

A node assigns values to its incident arcs when the values of the arcs satisfy a given condition. The following twelve types of primitive nodes are provided for coding the computation: *ditributor, conjunctor, selector, disjunctor, control recognizer, sink, get* node, *put* node, *constant* node, *data distributor, binary operation* node, and *unary operation* node. Their graphic symbols are shown in Figure 1.

When a control arrives at its unique input arc, after some delay, a *distributor* sends controls to all its output arcs. A *conjunctor* waits until all its input arcs have controls and then sends a control to its output arc. A *selector* sends controls to some of its arcs according to the value of its input data arc. That is, some of its output arcs are "selected" by the values of its input data. A *disjunctor* sends a control to its output arc after a control has arrived at one of its input arcs. A *control recognizer* identifies the input control arc at which a control has arrived, and produces the assigned number to the control arc to the output data arc. A *sink* is a sink of controls, which sends no control. A *get* node reads the contents of the memory address indicated by the input data arc. A *put* node writes the values of one of the input data arcs to the memory address indicated by the other input data arc. A *constant* node generates a constant value. A *data distributor* assigns the value of its input data arc to all its output data arcs. When a control arrives at its input control arc, after some delay, a *binary operation* node assigns the result of operation on the values of the two input data arcs to the output data arc and sends control to its output control arc. A *unary operation* node is defined similarly to a binary operation node.

All of the primitive C-graph elements explained above can be automatically translated into hardware elements. For example, a control arc is translated into a 1-bit register, which is set by the source node and is reset by the target node. The rules for translating the primitive nodes into DDL are given in Figure 2.

2.3 Examples

This section presents some practical examples of C-graphs.

Semaphores

Semaphores are primitive facilities for mutual exclusion[6]. Figure 3 shows the C-graph description for semaphores, which are an implementation of a "mutex" of Petri nets.

Loops

Figure 4 shows the n times iterations. The nodes labeled "counter," "counter: = 0" and "counter: = " are not primitive nodes, but can be easily defined. The node labeled "gen" starts to produce a sequence of values when a control arrives at the arc labeled "init." It then produces values each time a control arrives at the arc labeled "next." Figure 5 gives an example of a node that produces the sequence, 1, 2, ...

Pipelines

The C-graph description in Figure 4 represents sequential processing. Each execution starts after the previous one completes. If the body of the execution part holds conditions, the body can be exeucuted by pipelining. Pipelining is realized by producing a value before the previous execution completes. Figure 6 shows the C-graph description to realize pipelining.

Guard

Some parts of C-graphs must to be guarded to prevent the entrance of another control until the current execution completes. Some examples are the loops in Figure 4 and the pipelining in Figure 6. Figure 7 shows how the guard is realized. This facility is also implicitly used in Figures 3 and 6.

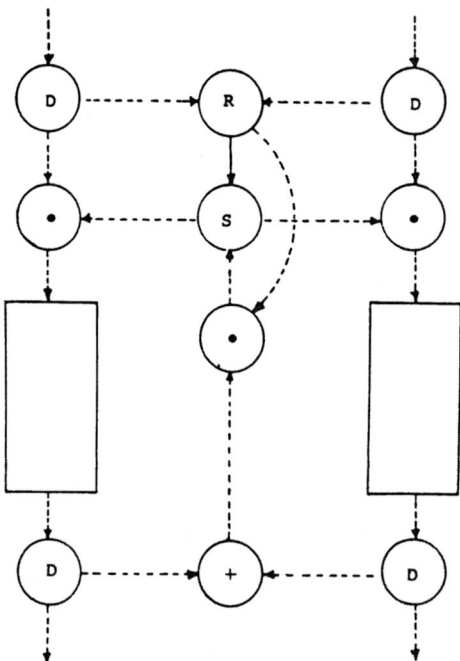

Figure 3. Semaphore

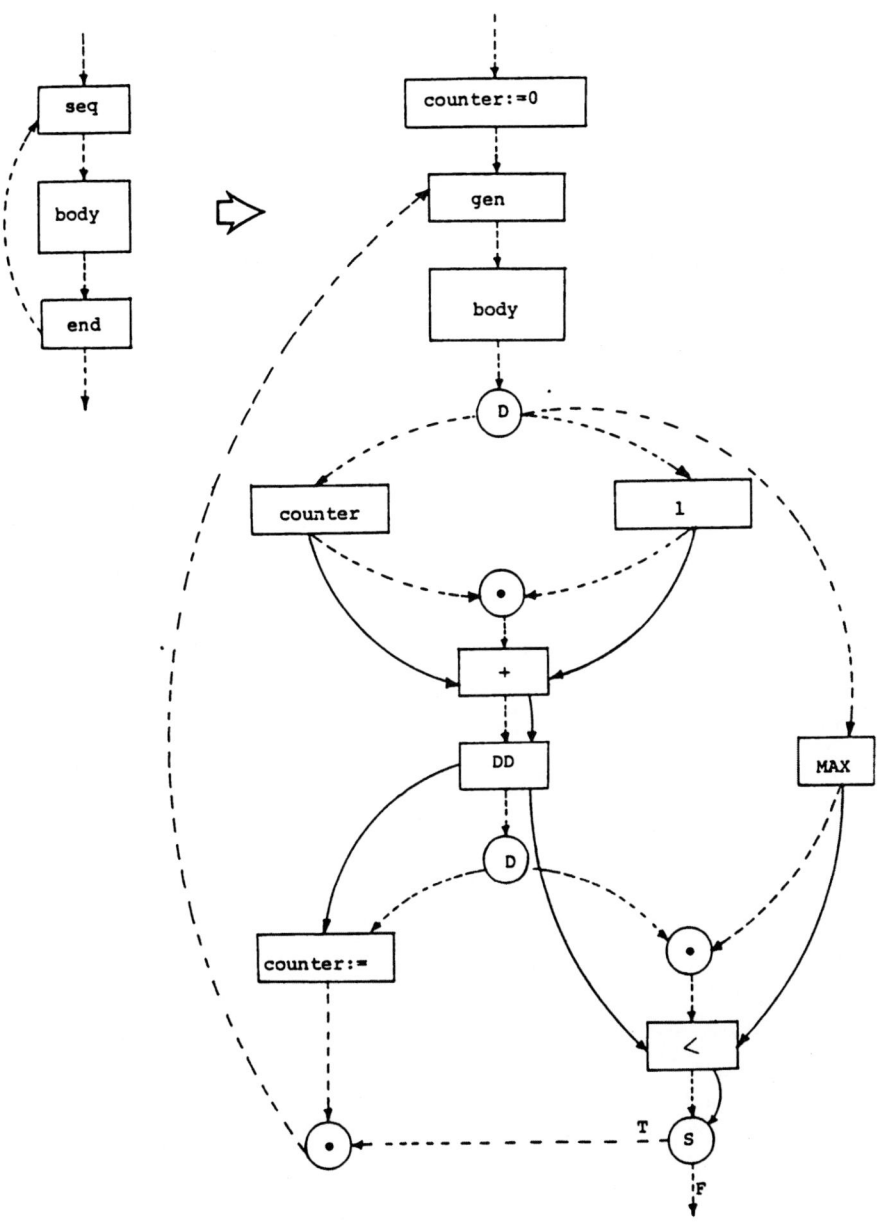

Figure 4. Loop

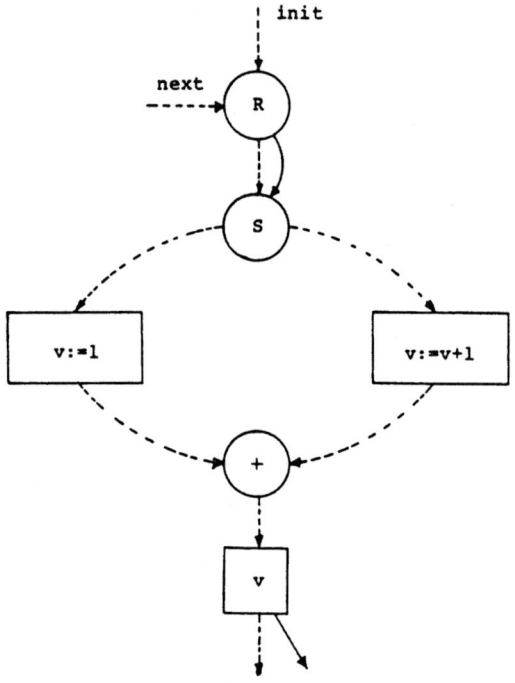

Figure 5. Definition of "gen"

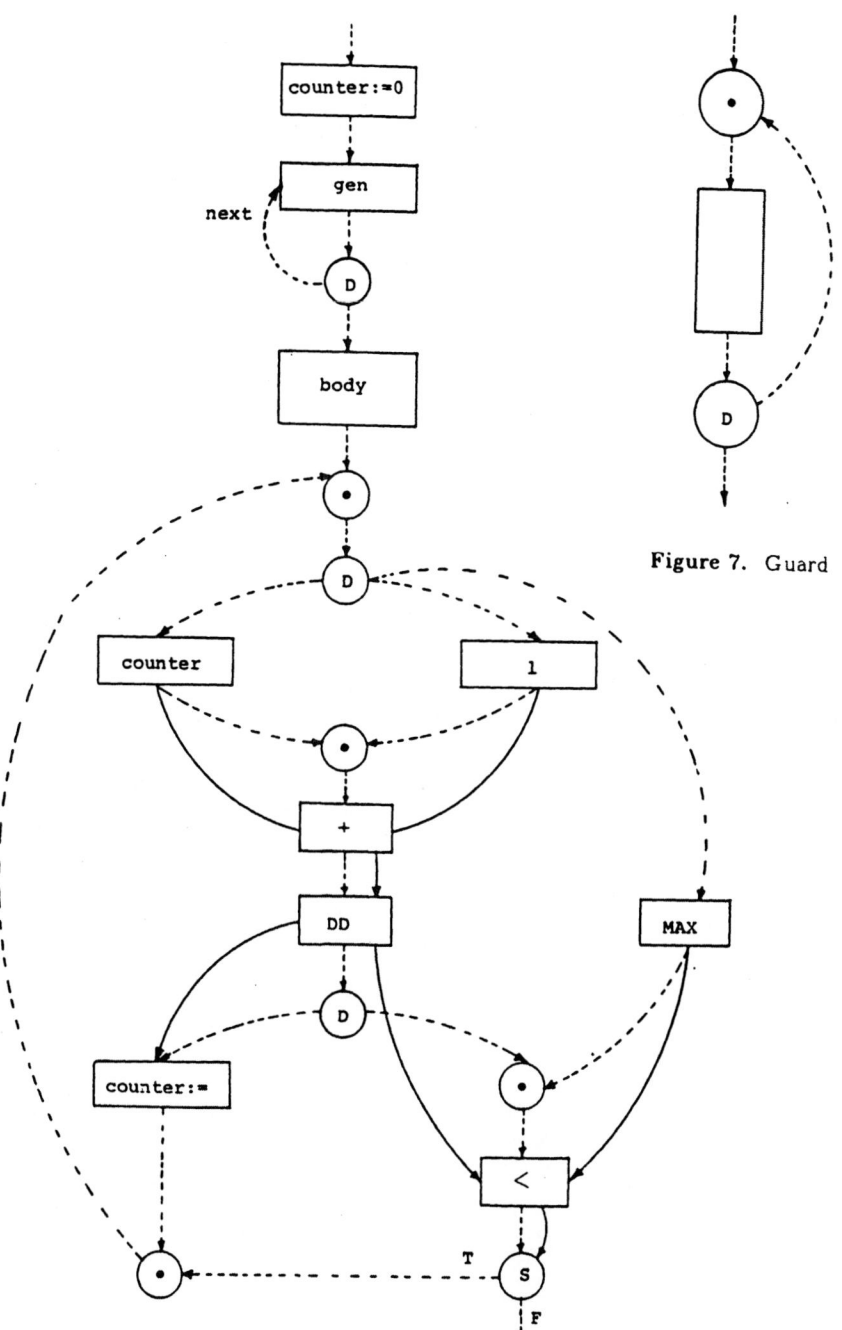

Figure 7. Guard

Figure 6. Pipelining

3. Parallel Computation Description Language

This section introduces the Parallel Computation Description Language (PCDL), which is used to describe a specific class of parallelism and whose syntax is based on that of Pascal[7] and Modula[8]. This section explains key PCDL facilities and rules to translate PCDL notations into C-graphs. Appendix B provides details on the entire PCDL syntax. PCDL is originally proposed by the authors[9]. In this paper, several new concepts are introduced to PCDL, and the syntax is modified slightly to clarify serialization and to provide a more convenient method of program coding. The syntax and data types are a minimal set for discuss partitioning and serialization in the next section. The syntax is described using extended BNF notation, which is also used for describing the syntax of Modula.

3.1 Facilities of PCDL

Types

PCDL supports standard types and an array type. Standard types include integer, real and Boolean types.

Variables

Variables are storages of values. In C-graphs, they are represented by an address in common memory and by data arcs. If data arcs are used, standard type variables and indexed array type variables are represented by primitive data arcs; array type variables are represented by a set of data arcs.

Functions and procedures

Because functions and procedures are handled as macros, function calls and procedure calls are treated as macro expansions. With this arrangement, recursive call is inhibited. A structure type is allowed for the result type of a function; that is, an array type is the current usage. Figure 8 shows the prototype of a C-graph representation for a function call.

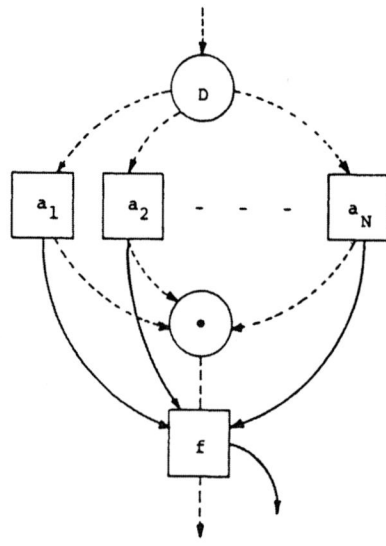

Figure 8. Function call
$f(a_1, a_2, \cdots, a_N)$

Parallel expressions and expression sequences

A parallel_expression and an expression_sequence represent a set of expressions, which are evaluated in parallel.

      parallel_expression = expression_set { "**cup**" expression_set }.
      expression_set = "{" expression_group ["|" condition] "}".
      expression_sequence = expression_vector { "|" expression_vector }.
      expression_vector = "[" expression_group ["|" condition] "]".
      expression_group = expression { "," expression }.
      condition = dummy_declaration { ";" dummy_declaration } { ";" constraint }.
      dummy_declaration = dummy_ident { "," dummy_ident } ":" constant ".." constant.
      dummy_ident = ident.

The expressions in the expression_group are evaluated in parallel. In the parallel_expression, the order of values is not specified; that is, the expression is considered a set of the values. The values in the expression_sequence, however, are ordered. The operator **cup** represents the union of sets of values, and the operator | represents the concatenation of the sequences of values. The operator **cup** is commutative, but the | is not. The dummies are the parameters of the expression_group. In the expression sequence, the dummy declarations specify the order of the values. For example, the dummy declaration "i,j: 1..2" represents a sequence of (i,j), such as (1,1), (1,2), (2,1) and (2,2). If the expression in the expression_group does not contain the dummies, the expression is copied for each set of values for the length of the sequence. The constraint is a Boolean expression, which constrain the range of the values of dummies. Parallel_expressions and expression_sequence appear in reduc_expressions and for_statements, which are explained in the next section.

Examples:
1. { x/y, m **mod** } is equivalent to { x **mod** y, x/y }.
2. [x/y, m **mod**] is not equivalent to [x **mod** y, x/y].
3. { A[i] | i: 1..4 } is equivalent to { A[1], A[2], A[3], A[4] }.
4. [0 | i: 1..3] is equivalent to [0, 0, 0].
5. { A[i] | i: 1..N; odd(i) } **cup** { A[i] | i: 1..N; **not** odd(i) } is equivalent to
   { A[i] | i: 1..N }.
6. { A[i,k]*B[k,j] | k: 1..N }.

Reduc_expressions

The reduc_expression specifies that an operator or a function is applied over all elements of the parallel_expression and expression_sequence. The effect is the same as the reduction operator in APL[10].

      reduc_expression = parallel_reduc | sequential_reduc.
      parallel_reduc = "**reduc**" reduc_operator parallel_expression.
      sequential_reduc = "**reduc**" reduc_operator expression_sequence.
      reduc_operator = add_operator | mul_operator | func_ident.
      add_operator = " + " | " - " | "or".
      mul_operator = "*" | "/" | "**div**" | "**mod**" | "**and**".

The reduc_operator is a binary operator or a function identifier which denotes a function with two identical type formal parameters. Because the result must be the same type as the operands, relational operators are not included in the reduc_operators. Because the

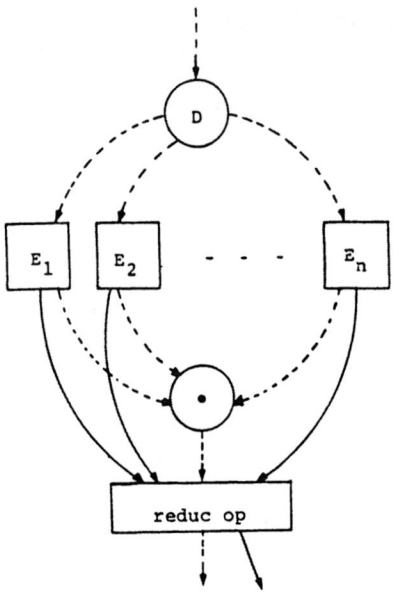

Figure 9(a). Parallel reduc

$$\textbf{reduc}(\mathrm{op},\{E_1,E_2,\cdots,E_N\})$$

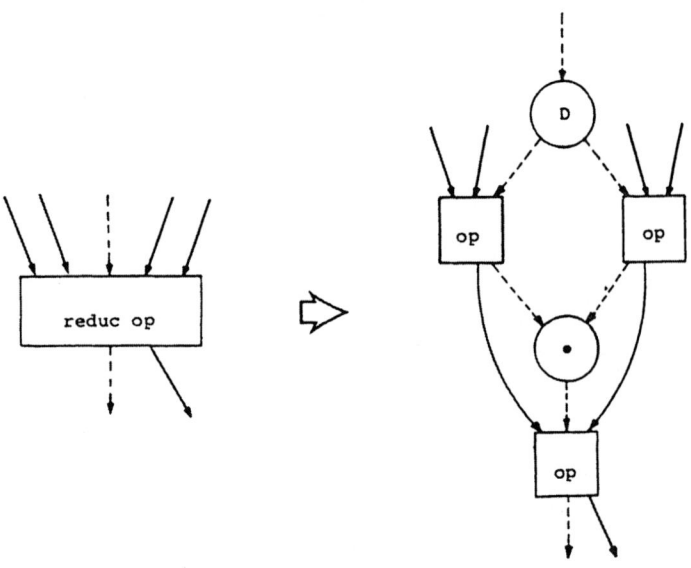

Figure 9(b). Definition of "reduc op"

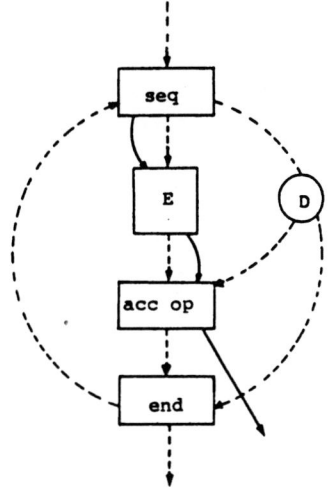

Figure 10(a). Sequential reduc
reduc(op, [E | s])

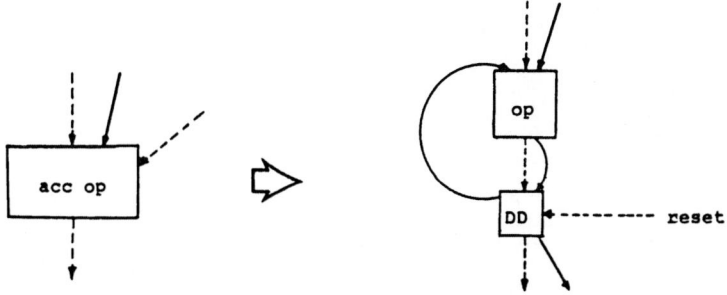

Figure 10(b). Definition of "acc op"

evaluation order is not specified in the parallel_reduc, the reduc_operator used in a parallel_reduc must be commutative and associative. In the sequential_reduc, the evaluations are performed according to the order of values of the expression_sequence. A parallel_reduc is translated into the C-graphs shown in Figure 9, the sequential_reduc is translated into the one shown in Figure 10.

Example:

The expressions, **reduc**(+ , { A[i] | i: 1..4}) and **reduc**(+ , [A[i] | i: 1..4]), produce the same value as A[1] + A[2] + A[3] + A[4]. In the first expression, the order of evaluations is not specified. For example, A[1] + A[2] and A[3] + A[4] are evaluated in parallel, then the two results are added. In the second expression, the order is from 1 to 4.

Parallel statements

A parallel_statement specifies that statments in the statement_sequence are executed in parallel.

parallel_statement = "{" statement_sequence ["|" condition] "}".

Expansions by dummy_variables are similar to those of expression_sets and expression_vectors. The translation rule from a parallel_statement to a C-graph is shown in Figure 11.

For_statements

The for_statement specifies that the statement_sequence between **do** and **od** be executed a specified number of times. The specified number is the cardinality of the range of the dummy placed next to the symbol **for**.

for_statement = parallel_for_statement | sequential_for_statement.
parallel_for_statement =
    "**for**" dummy_ident "**in**" parallel_expression
      "**do**" statement_sequence "**od**".
sequential_for_statement =
    "**for**" dummy_ident "**in**" expression_sequence
      "**do**" statement_sequence "**od**".

In the parallel_for_statement, the execution of the statement_sequence is performed asynchronously in parallel. The order of execution is not specified. If multiple statements assign the same variable, the result is not guaranteed. In the sequential_for_statement, the statement_sequences are executed in the order of the values in the expression_sequence.

A parallel_for_statement is translated to a parallel_statement before it is translated into a C-graph, so the C-graph derived from a parallel_for_statement is identical to that derived from a parallel_statement (Figure 11). A sequential_for_statement is translated into a C-graph in two ways. Figure 12 shows the usual sequential processing. The "seq" and "end" nodes are defined as shown in Figure 4. If pipelining is available, the sequential_for_statements are translated into the C-graphs shown in Figure 6.

Examples:

      for i **in** { x | x: 1..N } **do**
       for j **in** { y | y: 1..N } **do**
        C[i,j] := **reduc**(+ , { A[i,k]*B[k,j] | k: 1..N }) **od od**

represents that the N^2 assignments are executed in parallel and that N^3 multiplications are evaluated in parallel.

146

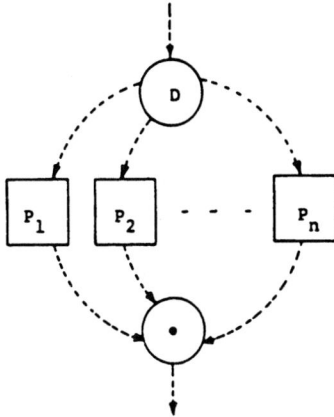

Figure 11. Parallel statement
$\{P_1; P_2; \cdots ; P_N\}$

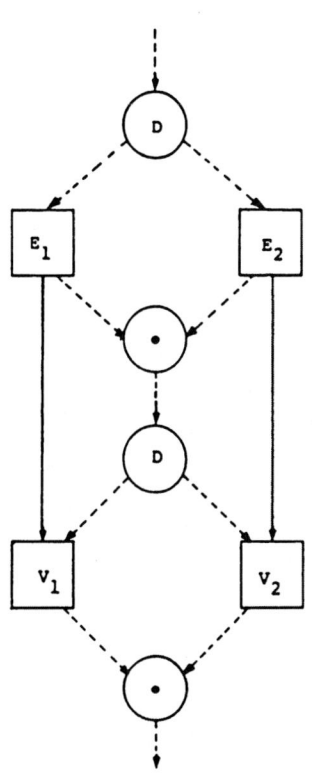

Figure 13. Multiple assignment
$[V_1, V_2] := [E_1, E_2]$

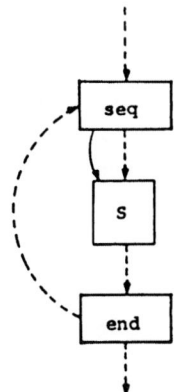

Figure 12. Sequential **for** statement
for v **in** [seq] **do** S **od**

147

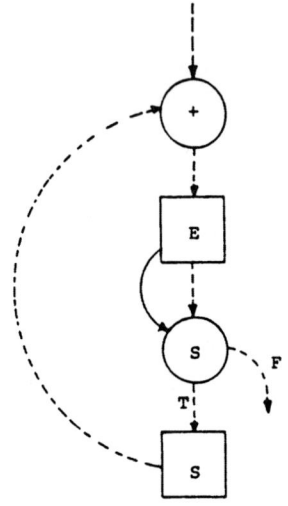

Figure 15. **while** statement
while E **do** S **od**

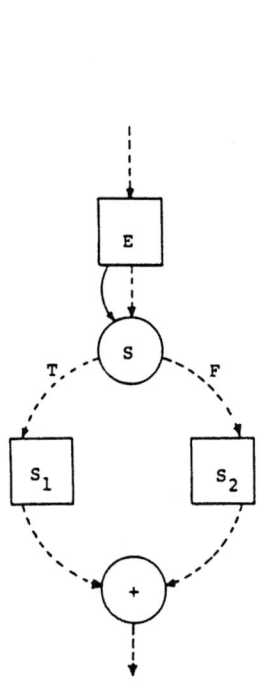

Figure 14. **if** statement
if E **then** S_1 **else** S_2 **fi**

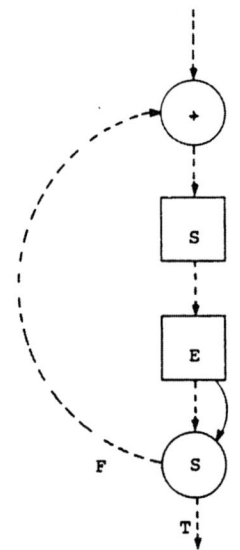

Figure 16. **repeat** statement
repeat S **until** E

Multiple assignments

The multiple assignment specifies that each value of the expression be assigned to a corresponding variable after all evaluations of the expression_sequence.

> multiple_assignment = variable_sequence ":=" expression_sequence.
> variable_sequence = variable_vector { "|" variable_vector }.
> variable_vector = "[" variable_group ["|" condition] "]".

The variable sequence and the expression sequence must have the same length, and the variable must have the same type as the corresponding expression. Figure 13 shows an example of a translation from a multiple assignment to a C-graph.

Example:
1. [x, y] := [y, x] exchanges the value of x and y.
2. [A[i] | i: 1..N – 1] := [A[i] | i: 2..N] specifies to shift the values of array A.

Other statements

The if_statement, while_statement, repeat_statement and case_statement are provided for sequential programming. Their semantics are identical to those of Modula. Translations of the if_statement, while_statement and repeat_statement into C-graphs are shown in Figures 14, 15 and 16, respectively.

3.2 Examples

Two practical examples in this section demonstrate the usefulness of PCDL to code parallel computation. Example 1 demonstrates three dimensional transformations and Example 2 demonstrates a sorting algorithm.

Example 1: Three Dimensional Transformations

The example of three dimensional transformation that follows shows the PCDL procedure to return the transformation matrix of the rotation about an arbitrary axis through an arbitrary point. Let us define (x,y,z) as a point through which the rotation axis passes, and (a,b,c) as the direction cosine of the axis. The rotation matrix R through an angle θ about this axis is represented as follows[11]:

$$R = T\, R_1 R_2 R_\theta R_2^{-1} R_1^{-1} T^{-1}$$

where

$$T = \begin{bmatrix} 1 & 0 & 0 & 0 \\ 0 & 1 & 0 & 0 \\ 0 & 0 & 1 & 0 \\ -x & -y & -z & 1 \end{bmatrix} \quad R_1 = \begin{bmatrix} 1 & 0 & 0 & 0 \\ 0 & c/v & b/v & 0 \\ 0 & -b/v & c/v & 0 \\ 0 & 0 & 0 & 1 \end{bmatrix},$$

$$R_2 = \begin{bmatrix} v & 0 & a & 0 \\ 0 & 1 & 0 & 0 \\ -a & 0 & v & 0 \\ 0 & 0 & 0 & 1 \end{bmatrix}, \quad R_\theta = \begin{bmatrix} \cos\theta & -\sin\theta & 0 & 0 \\ \sin\theta & \cos\theta & 0 & 0 \\ 0 & 0 & 1 & 0 \\ 0 & 0 & 0 & 1 \end{bmatrix},$$

$$v = (b^2 + c^2)^{1/2}$$

Figure 17 shows the program of this computation.

Example 2: Sorting by counting

Because sorting is important for practical application and, at the same time, presents some interesting theoretical problems, numerous sorting algorithms have been devised. The sorting problem can be formulated as follows[12]: We are given a sequence of n elements $d_1, d_2, ..., d_n$ drawn from a set having a linear order, which we shall usually denote $\le$. We are to find a permutation p of these n elements that will map the given sequence into a non-

```
type trans = array[1..4,1..4] of real;

function unit: trans;
   var T: trans;
begin
   [ T[i,j] | i,j: 1..4 ] := [ 1, 0, 0, 0,  0, 1, 0, 0,  0, 0, 1, 0,  0, 0, 0, 1 ];
   unit: = T
end unit;

function translate(x,y,z: real): trans;
   var T: trans;
begin
   T: = unit;
   { T[4,1]: = x;  T[4,2]: = y;  T[4,3]: = z };
   translate: = T
end translate;

function rotate(c,s: real; axis: integer): trans;
(* axis: 1 → x, 2 → y, 3 → z *)
   var T: trans;
       i,j: integer;
begin
   T: = unit;
   { i: = axis mod 3 + 1;  j: = (axis + 1) mod 3 + 1 };
   { T[i,i]: = c;  T[i,j]: = − s;  T[j,i]: = s;  T[j,j]: = c };
   rotate: = T
end rotate;

function mult(A,B: trans): trans;
   var C: trans;
begin
   for i in [ x | x: 1..4 ] do
      for j in [ y | y: 1..4 ] do
         C[i,j] := reduc( +, { A[i,k]*B[k,j] | k: 1..4 })
      od
   od;
   mult: = C
end mult;

function rotation(x,y,z,a,b,c,th: real): trans;
   var v: real;
begin
   v := sqrt(sqr(b) + sqr(c));
   rotation :=
      mult(translate( − x, − y, − z), mult(rotate(c/v, − b/v,1),
      mult(rotate(v,a,2), mult(rotate(cos(th),sin(th),3),
      mult(rotate(v, − a,2), mult(rotate(c/v,b/v,1), translate(x,y,z)))))))
end rotation;
```

Figure 17. Three Dimensional Transformation

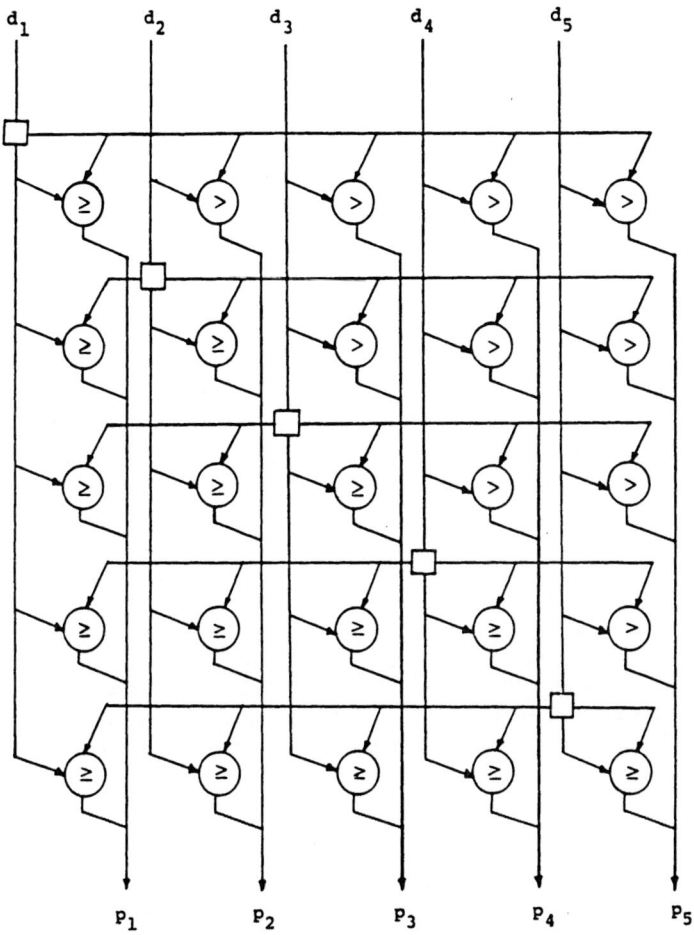

Figure 18. Sorting by counting

decreasing sequence $d_{p(1)}, d_{p(2)}, ..., d_{p(n)}$ such that $d_{p(i)} \leq d_{p(j)}$ for $1 \leq i \leq j \leq n$. Usually we produce the sorted sequence itself rather than the sorting permutation p.

One method of computing the permutation p is to count the number of elements d_j such that $d_j \leq d_i$. The sorting algorithm using this method is called "sorting by counting." The PCDL code to represent the algorithm is as follows:

```
for i in { x | x: 1..N } do
    p[i] := reduc( + ,{ if d[i] > d[j] then 1 else 0 fi | j: 1..N }) od
```

In the above code, we assume that the values are different from one another. If some elements have the same value, it is necessary to use the different comparisons between $i \geq j$ and $i < j$.

```
for i in { x | x: 1..N } do
    p[i] := reduc( + , { if d[i]  d[j] then 1 else 0 fi | j: 1..N; ij { cup
                       { if d[i] > d[j] then 1 else 0 fi ; j: 1..N; i<j }) od
```

The C-graph representation of this statement is shown in Figure 18. The control arcs are omitted to make the graph easier to read.

After the permutation p is determined, the exchange of the values is coded as follows:

$$[\ d[p[i]] \ | \ i: 1..N \ ] := [\ d[i] \ | \ i: 1..N \ ]$$

4. Partitioning and Serialization

This section discusses how PCDL can be applied for partitioning and serialization. Partitioning and serialization are fundamental techniques used in designing parallel processor systems having many processors. But although PCDL has proved a useful tool for utilizing those techniques, we have not yet formed the answer to the following question:

Can these techniques be applied automatically?

This problem will be the focus of our future work.

4.1 Partitioning

The statements in a parallel_for_statement should be independent of one another. If they have interdependence, they cannot be executed in parallel without intercommunication. There is the following well-known conditions for deciding whether two statements s_i and S_j can be executed in parallel[13]. Let O_i and I_i be the set of output variables and input variables of the statement S_i, respectively.

$$(*) \quad I_i \cap O_j = \phi, \quad O_i \cap I_j = \phi, \quad O_i \cap O_j = \phi$$

These conditions $(*)$ are called parallel processable conditions.

Statements in a parallel_for_statement do not always satisfy the condition $(*)$. But in some cases, the statements can be divided into several classes so as to satisfy the condition. The following example of a simple sorting algorithm shows how this is done. The basic idea of the algorithm is as follows:

1. While an index i is found such that $a_i > a_{i+1}$ exchange the values of a_i and a_{i+1}.
2. If such index is not found, sort the sequence.

The next program is a sample of this algorithm in PCDL.

```
repeat for i in { x | x: 1..N - 1 } do
    if a[i] > a[i + 1] then [ a[i], a[i + 1] ] := [ a[i + 1], a[i] ] fi od
until reduc(and, { a[i] >= a[i + 1] | i: 1..N - 1 })
```

The program is faulty because, if the condition $a[i - 1] > a[i] > a[i + 1]$ is satisfied for some

$2 \leq i \leq N - 1$, two statements will try to assign the value to $a[i]$. By dividing the set of statements into two classes so that two statements in different classes are not executed in parallel, we obtain the following correct program.

> **repeat for** i **in** { x | x: 1..N − 1; odd(i) } **do**
>   **if** $a[i] > a[i + 1]$ **then** [$a[i]$, $a[i + 1]$] := [$a[i + 1]$, $a[i]$] **fi od**;
>   **for** i **in** { x | x: 1..N − 1; **not** odd(i) } **do**
>   **if** $a[i] > a[i + 1]$ **then** [$a[i]$, $a[i + 1]$] := [$a[i + 1]$, $a[i]$] **fi od**
> **until reduc(and**, { $a[i] >= a[i + 1]$ | i: 1..N − 1 })

4.2 Serialization

This section discusses the problem of serialization. Serialization is a process to replace parallel operations with sequential ones. Even if chips containing many processors can be fabricated by VLSI technology, the amount of hardware resources is not infinite. It is, therefore, very important to find an easy way to reduce hardware.

Serialization of C-graphs consists of two steps. The first step is to find the set of isomorphic subgraphs of the C-graph which are given semantics of similar operations. The second step of the serialization is to build the general subgraph, which has the semantics to perform all the operations with the mechanism for the sequential execution of each operation. If each operation consists of several steps, the control mechanism can be made to execute the operations using the pipelining method. The first step of serialization on a non-restricted C-graph is a time-consuming operation and, in most cases, not feasible (see Appendix C). When a computation is written in PCDL, the statements in the for_statements and parallel_statements can be translated to the general subgraph explained above, and the serialization of parts of computations can be performed in a clear and simple way.

The following presents the transformation rules of the PCDL statements that can be used for serialization. Some examples of serialization are also given.

Transformation rules

Nine transformation rules are provided for serialization. To be exact, the rules TR4 and TR7 are serialization, and the others are equivalent transformations that are used to transform descriptions into the form to which rules TR4 and TR7 can be applied. Applying these rules is all that is required to achieve systematic serialization. The second example in the following section shows that, in most cases, serialization presents many choices.

> (TR1) {P_1} **cup** {P_2} **cup** ... **cup** {P_n}
>   ↓   ↑
> {$P_{p(1)}$} **cup** {$P_{p(2)}$} **cup** ... **cup** {$P_{p(n)}$}
> where p is a permutation function.

> (TR2) **for** v **in** {P_1} **cup** {P_2} **cup** ... **cup** {P_n} **do** S **od**
>   ↓   ↑
> { **for** v **in** {P_1} **do** S **od**;
>   **for** v **in** {P_2} **cup** {P_3} **cup** ... **cup** {P_n} **do** S **od** }

> (TR3) **for** v **in** [P_1] | [P_2] | ... | [P_n] **do** S **od**
>   ↓   ↑
> **for** v **in** [P_1] **do** S **od**;
> **for** v **in** [P_2] | [P_3] | ... | [P_n] **do** S **od**

(TR4) **for** v **in** $\{P_1\}$ **cup** $\{P_2\}$ **cup** ... **cup** $\{P_n\}$ **do** S **od**

      ↓   ↑*

    for v **in** $[P_1] | [P_2] | ... | [P_n]$ **do** S **od**

*) The statements parameterized by v such that

    $v \in \{P_1\}$ **cup** $\{P_2\}$ **cup** ... **cup** $\{P_n\}$

should satisfy the parallel computable conditions.

(TR5) $\{ S_1; S_2; ... ; S_n \}$

      ↓   ↑

    $\{ S_1 \}$ **cup** $\{ S_2; S_3; ... ; S_n \}$

(TR6) $\{ S_1 \}$ **cup** $\{ S_2; S_3; ... ; S_n \}$

      ↓   ↑

    $\{ S_1; \{ S_2; S_3; ... ; S_n \} \}$

(TR7) **reduc**(op, $\{P_1\}$ **cup** $\{P_2\}$ **cup** ... **cup** $\{P_n\}$)

      ↓   ↑**

    reduc(op, $[P_1] | [P_2] | ... | [P_n]$)

**) The operator, op, should be commutative and associative.

(TR8) **reduc**(op, $\{P_1\}$ **cup** $\{P_2\}$ **cup** ... **cup** $\{P_n\}$)

      ↓   ↑

    reduc(op, $\{$ **reduc**(op, $\{P_1\}$), **reduc**(op, $\{P_2\}$), ... , **reduc**(op, $\{P_n\}$)

(TR9) $[V(d) | C(d)] := [E(d) | C(d)]$

      ↓***   ↑

    for d **in** $\{ d | C(d) \}$ **do** $V(d) := E(d)$ **od**

***) The variables $V(d)$ must not be used in the expression $E(d)$. If the condition does not hold, the description should be transformed as follows:

    for d **in** $\{ d | C(d) \}$ **do** $U(d) := E(d)$ **od**;

    $V := U$;

where U is the same type variable as V.

4.3 Examples
Multiplication of matrices

The following is a simple example of how PCDL is used to code an $N \times N$ matrix multiplication.

    (1) $[C[i,j] | i,j: 1..N] := [$**reduc**$(+ , \{A[i,k]^*B[k,j] | k: 1..N\}) | i,j: 1..N]$

Transformation is possible using rules TR7 and TR9. Using rule TR7, the description is transformed as follows:

    (2) $[C[i,j] | i,j: 1..N] := [$**reduc**$(+ , [A[i,k]^*B[k,j] | k: 1..N]) | i,j: 1..N]$

This transformation is an example of serialization. Rule TR9 can be applied to (1) because the variable C is not used in the right side of the assignment symbol. The transformation produces the following code.

(3) **for** i **in** {x | x: 1..N} **do**
    for j **in** {y | y: 1..N} **do**
        C[i,j] := **reduc**(+ ,{A[i,k]•B[k,j] | k: 1..N}) **od od**

This transformation does not itself serialize the description, but this form can be serialized by applying rules TR4 and TR7. If all possible transformations are performed, the description is transformed into the following statements:

(4) **for** i **in** [x | x: 1..N] **do**
    for j **in** [y | y: 1..N] **do**
        C[i,j] := **reduc**(+ ,[A[i,k]•B[k,j] | k: 1..N]) **od od**

Pattern matching

As an example of more sophisticated serialization, we will discuss a pattern matching algorithm. Suppose a string $S=(s_1,s_2,\cdots,s_N)$ and a pattern $P=(p_1,p_2,\cdots,p_N)$ are given. Let us find all substrings of S that match P exactly. The output is a string of Boolean values $B=(b_1,b_2,\cdots,b_N)$ where b_i = true if $s_{i+j-1}=p_j$ for all j such that $1 \leq j \leq M$ and b_i = false otherwise. If N×M comparators are available, we can code this algorithm in PCDL as follows:

(5) [b[i] | i: 1..N] := [**reduc**(and,
    { **if** i+j-1 <= N **then** s[i+j-1]=p[j] **else** false **fi** | j: 1..M }) | i: 1..N]

This coding is illustrated in Figure 19. The code in (5) is equivalent to the code in (6) because of rule TR9.

(6) **for** i **in** { x | x: 1..N } **do**
    b[i] := **reduc**(and,
    { **if** i+j-1 <= N **then** s[i+j-1]=p[j] **else** false **fi** | j: 1..M }) **od**

This code can be serialized according to the variables i (rule TR4) and j (rule TR7) (Figure 20(a) and (b)). Figure 20(a) and (b) show the shared parts, which are represented by dotted lines.

The above methods are not the only serializations possible. For example, if the parallel_expression {x| x: 1..N} is split into two expression sets, such as

{x| x: 1..N; odd(x)} **cup** {x| x: 1..N; **not** odd(x)}

rule TR2 can be applied. If each statement is serialized, the code is transformed as follows:

(7) { **for** i **in** [x | x: 1..N; odd(x)] **do**
    b[i] := **reduc**(and,
    { **if** i+j-1 <= N **then** s[i+j-1]=p[j] **else** false **fi** | j: 1..M }) **od**;
    for i **in** [x | x: 1..N; **not** odd(x)] **do**
    b[i] := **reduc**(and,
    { **if** i+j-1 <= N **then** s[i+j-1]=p[j] **else** false **fi** | j: 1..M }) **od** }

In this implementation, 2×M comparators are used. We can get code with various degrees of complexity of resources by applying nine transformation rules.

Figure 20(c) shows another serialization. It seems as natural as Figure 20(a) and 20(b). But when this serialization is used, the description is as follows:

(8) **var** B: **array**[-M..N] **of** Boolean; V: **array**[1..M] **of** Boolean;
    for i **in** [x | x: 1..N] **do**
    [V[j] | j: 1..M] | [B[i-M+1]] :=
    [s[i]=p[1]] | [V[j-1] **and** (s[i]=p[j]) | j: 1..M] **od**

The transformation is somewhat artificial when coded in PCDL.

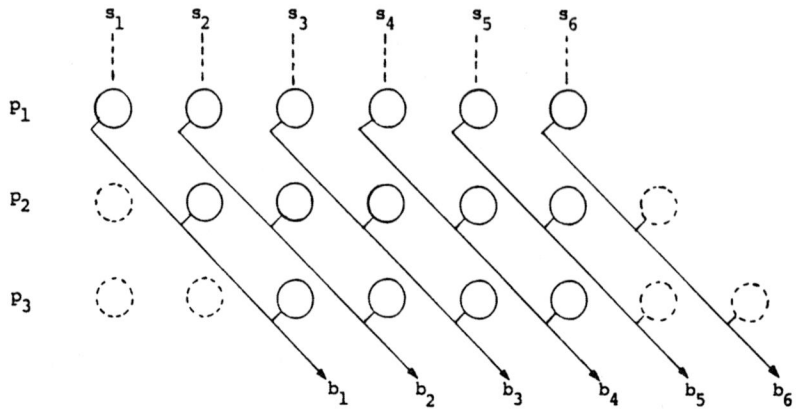

Figure 19. Pattern matching

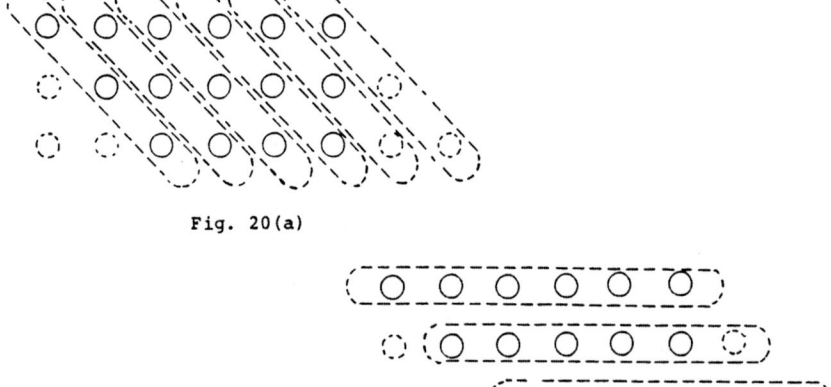

Fig. 20(a)

Fig. 20(b)

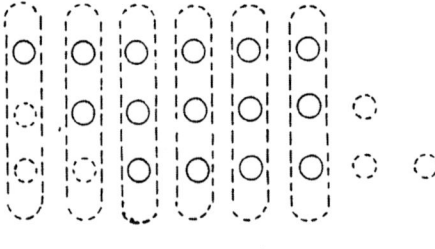

Fig. 20(c)

Figure 20. Three types of serializations

5. Conclusions

In this paper, we have defined C-graphs which represent computations on a general model of parallel computation. PCDL was defined as a user interface language for C-graphs, and can be used to code locally parallel computations into a compact code. We have also shown how to implement parallelism in the actual hardware. PCDL has proved to be a high level language powerful enough to code parallelism. Especially important, PCDL provides systematic serialization.

Acknowledgements

We would like to thank Professor Satoru Kawai, Kazunori Yamaguchi and Katsumi Kanasaki for their valuable comments and ideas.

Appendix A: The Formal Definition of C-graphs

Let C-arc and D-arc denote the set of control arcs and data arcs, respectively. The value domains of arcs are defined as follows:

$B=\{T, F\}$: truth value assigned to control arcs,

$V = \{Int, Float, Bool, Addr\}$: a set of domains: the values belonging to the domain may be assigned to data arcs. Int is integers, Float is floating point numbers, Bool is Boolean values, and Addr is addresses of memory.

Let $Time \subset Z$ denote the values which represent time, where Z denotes the integers. The following functions are provided for inspecting the value of arcs:

$$Dom: D\text{-arc} \rightarrow V, \quad Dom(x) \text{ is the domain of the value of } x,$$
$$c: C\text{-arc} \times Time \rightarrow B,$$
$$v: D\text{-arc} \times Time \rightarrow \sum_{V \in V} V$$
$$m: Addr \times V \times Time \rightarrow \sum_{V \in V} V$$

Notations

Let Node denote the set of nodes. A definition of nodes consists of a header and a body. The header gives information on the incident arcs and attributes of the nodes. "IN" indicates input arcs, "OUT" indicates output arcs, and "Delay" indicates the time necessary for the node to complete its operation. The body is in the form '$X \Rightarrow Y$' whose preceded quantifier ($\forall t \in Time$) is omitted. X is a triggering condition on which the node performs its operations, and Y is the result of the operations. The symbol "$\Rightarrow$" implies that if the condition is not satisfied, the node perform no operations.

Node = Distributor + Conjunctor + Selector + Disjunctor + Sink + ControlRecognizer
     + Get + Put + Const + DataDistributor + Add + ...

Distributor: IN: $a \in$ C-arc; OUT: $A \subset$ C-arc; Delay: $d \in Z$;
$$(\forall a \in A)(c(a,t) \land \neg c(x,t)) \implies (\forall x \in A)(\neg c(a,t+d) \land c(x,t+d)).$$

Conjunctor: IN: $A \subset$ C-arc; OUT: $a \in$ C-arc; Delay: $d \in Z$;
$$(\forall x \in A)(c(x,t) \land \neg c(a,t)) \implies (\forall x \in A)(\neg c(x,t+d) \land c(a,t+d)).$$

Selector: IN: $a \in$ C-arc, $b \in$ D-arc; OUT: $A \subset$ C-arc; Delay: $d \in \mathbf{Z}$;
$\quad$ Poset$((A, \leqq_1)) \wedge$ Poset$((\text{Dom}(b), \leqq_2)) \wedge c(a,t) \wedge (\forall x \in A) \neg c(x,t) \implies$
$\quad \neg c(a,t+d) \wedge (\forall x \in A)(\text{ord}(x) = \text{ord}(v(b,t)) \rightarrow c(x,t+d))$.
$\quad$ where Poset$((S, \leqq))$ is a predicate to examine whether S is a partially ordered set or
$\quad$ not, and ord: $V \rightarrow \mathbf{Z}$ is a function to give the order of a value.
$\qquad$ ord$(v) = |\{ x \mid x < v \}|$

Disjunctor: IN: $A \subset$ C-arc; OUT: $a \in$ C-arc; Delay: $d \in \mathbf{Z}$;
$\quad (\exists x \in A)(c(x,t) \wedge \neg c(a,t)) \implies (\forall x \in A)(c(x,t) \rightarrow \neg c(x,t-d)) \wedge c(a,t+d)$.

Sink: In: $a \in$ C-arc; Delay: $d \in \mathbf{Z}$;
$\quad c(a,t) \implies \neg c(a,t+d)$.

ControlRecognizer: IN: $A \subset$ C-arc; OUT: $a \in$ C-arc, $b \in$ D-arc; Delay: $d \in \mathbf{Z}$;
$\quad$ Poset$((A, \leqq)) \wedge (\exists x \in A)(c(x,t) \wedge \neg c(a,t)) \implies$
$\quad (c(x,t) \rightarrow \neg c(x,t+d)) \wedge c(a,t) \wedge v(b,t+d) = \text{ord}(x)$.

Get: IN: $a \in$ C-arc, $\text{loc} \in$ D-arc; OUT: $b \in$ C-arc, $\text{val} \in$ D-arc; Delay: $d \in \mathbf{Z}$;
$\quad \text{Dom}(\text{loc}) = \text{Addr} \wedge c(a,t) \wedge \neg c(b,t) \implies$
$\quad \neg c(a,t+d) \wedge c(b,t+d) \wedge v(\text{val}, t+d) = m(v(\text{loc},t), \text{Dom}(\text{val}), t)$.

Put: IN: $a \in$ C-arc, loc, $\text{val} \in$ D-arc; OUT: $b \in$ C-arc; Delay: $d \in \mathbf{Z}$;
$\quad \text{Dom}(\text{loc}) = \text{Addr} \wedge c(a,t) \wedge \neg c(b,t) \implies$
$\quad \neg c(a,t+d) \wedge c(b,t) \wedge m(v(\text{loc},t), \text{Dom}(\text{val}), t+d) = v(\text{val},t)$.

Const: IN: $a \in$ C-arc; OUT: $b \in$ C-arc, $\text{val} \in$ D-arc; Delay: $d \in \mathbf{Z}$; Value: $\text{value} \in \underset{v \in V}{+} V$;
$\quad \text{value} \in \text{Dom}(\text{val}) \wedge c(a,t) \wedge \neg c(b,t) \implies \neg c(a,t+d) \wedge c(b,t+d) \wedge v(\text{val},t+d) = \text{value}$.

DataDistributor: IN: $a \in$ C-arc, $x \in$ D-arc; OUT: $b \in$ C-arc, $X \subset$ D-arc; Delay: $d \in \mathbf{Z}$;
$\quad c(a,t) \wedge \neg c(b,t) \implies (\forall y \in X)(\neg c(a,t+d) \wedge c(b,t+d) \wedge v(y,t+d) = v(x,t))$.

Add: IN: $a \in$ C-arc, x,$y \in$ D-arc; OUT: $b \in$ C-arc, $z \in$ D-arc; Delay $d \in \mathbf{Z}$;
$\quad \text{Dom}(x) = \text{Dom}(y) = \text{Dom}(z) \wedge (\text{Dom}(x) = \text{Int} \vee \text{Dom}(x) = \text{Float}) \wedge c(a,t) \wedge \neg c(b,t) \implies$
$\quad \neg c(a,t+d) \wedge c(b,t+d) \wedge v(z,t+d) = v(x,t) + v(y,t)$.

Appendix B: Syntax of PCDL

```
constant_declaration = const_ident " = " constant.
constant_ident = ident.
constant = unsigned_constant | ( " + " | " - " ) number.
unsigned_constant = const_ident | number.

type_declaration = type_ident " = " type.
type_ident = ident.
type = type_ident | array_structure.
array_structure = "array" "[" index_range_list "]" "of" type.
index_range_list = index_range { "," index_range }.
index_range = constant ".." constant.

variable_declaration = var_ident_list ":" type.
var_ident_list = var_ident { "," var_ident }.
var_ident = ident.
variable = var_ident | func_ident | variable "[" indices "]".
```

```
indices = expression } "," expression {.

expression = simple_expression [ rel_operator simple_expression ].
rel_operator = "=" | "< >" | "<=" | "<" | ">" | ">=".
simple_expression = [ "+" | "-" ] term } add_operator term {.
add_operator = "+" | "-" | "or".
term = factor } mul_operator factor {.
mul_operator = "•" | "/" | "div" | "mod" | "and".
factor = unsigned_constant | variable | function_call | "(" expression ")" | "not" factor
         | dummy_ident | if_expression | reduc_expression.
function_call = func_ident "(" [ parameterlist ] ")".
if_expression = "if" expression "then" expression
        } "elif" expression "then" expression { "else" expression "fi".
reduc_expression = parallel_reduc | sequential_reduc.
parallel_reduc = "reduc" "(" reduc_operator parallel_expression ")".
sequential_reduc = "reduc" "(" reduc_operator expression_sequence ")".
reduc_operator = add_operator | mul_operator | func_ident.
parallel_expression = expression_set } "cup" expression_set {.
expression_set = "}" expression_group [ "|" condition ] "{".
expression_sequence = expression_vector } "|" expression_vector {.
expression_vector = "[" expression_group [ "|" condition ] "]".
expression_group = expression } "," expression {.
condition = dummy_declaration } ";" dummy_declaration { } ";" constraint {.
dummy_declaration = dummy_ident } "," dummy_ident { ":" constant ".." constant.
dummy_ident = ident.
constraint = Boolean_expression } "," Boolean_expression {.
Boolean_expression = expression.

statement = assignment | procedure_call | if_statement | case_statement
        | while_statement | repreat_statement | for_statement | parallel_statement.
assignment = simple_assignment | multiple_assignment.
simple_assignment = variable ": =" expression.
multiple_assignment = variable_sequence ": =" expression_sequence.
variable_sequence = "[" variable_group [ "|" condition ] "]".
variable_group = variable } "," variable {.
procedure_call = proc_ident [ "(" parameter_list ")" ].
parameter_list = parameter } "," parameter {.
parameter = expression | variable.
statement_sequence = statement } ";" statement {.
if_statement = "if" expression "then" statement_sequence
        } "elif" expression "then" statement_sequence {
        [ "else" statement_sequence ] "fi".
case_statement = "case" expression "of" case } "," case { "esac".
case = [case_labels ":" "begin" statement_sequence "end"].
case_labels = constant } "," constant {.
while_statement = "while" expression "do" statement_sequence "od".
repeat_statement = "repeat" statement_sequence "until" expression.
for_statement = parallel_for_statement | sequential_for_statement.
parallel_for_statement =
```

"**for**" dummy_ident "**in**" parallel_expression "**do**" statement_sequence "**od**".
sequential_for_statement =
    "**for**" dummy_ident "**in**" expression_sequence "**do**" statement_sequence "**od**".
parallel_statement = "}" statement_sequence ["▫" condition] "{".

procedure_declaration =
    "**procedure**" proc_ident ["(" formal_parameters ")"] ";" block proc_ident.
proc_ident = ident.

function_declaration =
    "**function**" func_ident ["(" formal_parameters ")"] ":" type_ident ";" block func_ident.
func_ident = ident.

formal_parameters = section { ";" section }.
section = ["**var**"] var_ident { "," var_ident } ":" formal_type.
formal_type = type_ident.
block = { declaration_part } [statement_part]
declaration_part = "**const**" { constant_declaration ";" } | "**type**" } type_declaration ";" }
    | "**var**" { variable_declaration ";" } | procedure_declaration ";"
    | function_declaration ";".
statement_part = "**begin**" statement_sequence.

program = block "end".

Appendix C: Algorithm to Find Equivalent Subgraphs

Definition: Two directed graphs, $G_1 = (N_1, A_1, f_1)$ and $G_2 = (N_2, A_2, f_2)$, are *equivalent* according to a semantic function s defined on N_1, N_2, A_1 and A_2, if there are two isomorphisms, $h_N: N_1 \rightarrow N_2$ and $h_A: A_1 \rightarrow A_2$, where $s(n) = s(h_N(n))$ for all $n \in N_1$, and $s(a) = s(h_A(a))$ for $\forall a \in A_1$.

Input: A directed graph $G = (N, A, f)$ and a connected directed graph $G_0 = (N_0, A_0, f_0)$. And a semantic function s defined on N, A, N_0 and A_0.

Output: A set of equivalent subgraphs of G.

Method: We begin making a spanning tree of G_0, $T_0 = (NT_0, AT_0, fT_0)$.
    Let r_0 be the root of T_0.
    $X := \phi$;
    for each $n \in N$ where $s(n) = s(r_0)$ do
        Let $T = (NT, AT, fT)$ be a subgraph of G which is a tree with node n as its root and is equivalent to T_0.
        Let tn be an isomorphism between NT and NT_0.
        Let ta be an isomorphism between AT and AT_0.
        If we can construct a 1 to 1 mapping $h_A: A_0 \rightarrow A$ such that
            $h_A|_{AT_0} = ta^{-1}$,
            $s(a_0) = s(h_A(a_0))$ for $\forall a_0 \in A$, and
            $f(h(a_0)) = (tn^{-1}(n_1), tn^{-1}(n_2))$ if $f_0(a_0) = (n_1, n_2)$,
        then
            /* $G_1 = (NT, h(A_0), f|_{h_A(A_0)})$ is equivalent to G_0. */
            $X := X \cup \{G_1\}$.   □

References

[1] Peterson,J.L., "Petri Nets," *ACM Computer Surveys* **9**, 3 (Sept. 1977), 223-252.

[2] J.R.McGraw, "Data Flow Computing: Software Development," *Proc. the 1st International Ionference on Distributed Computing Systems,* Oct. 1979, 242-251.

[3] J.B.Dennis, "The Varieties of Data Flow Computers," *Proc. the 1st International Conference on Distributed Computing Systems,* Oct. 1979, 430-439.

[4] C.G.Bell and A.Newell, *Computer Structures: Reading and Examples,* McGraw-Hill, 1971.

[5] Dietmeyer,D.L. & Duly,J.R., "Register Transfer Language and thier Translation," *Digital System Design Automation: Languages, Simulation & Data Base,* Computer Science Press, 1975, 117-218

[6] Dijkstra,E.W., "Cooperating sequential process," *Programming Languages,* F.Gennys, ed., Academic Press, 1968.

[7] Jensen,K. and Wirth,N. *PASCAL User Manual and Report,* second edition, Spring-Verlag, 1978.

[8] Wirth,N., "Modula: a Language for Modular Multiprogramming," *Software-Practice and Experience,* **7**, 1 (Jan. 1977), 3-35.

[9] Iizawa,A., Kunii,T.L. and Kawaii,S., "A Graph-based Hardware Design Specification System," *Proc. Fifteenth Hawaii International Conference on System Science,* (Jan. 1982), 122-131.

[10] *APL Language,* GC26-3847-4, IBM, 1978.

[11] Newman,W.M. and Sproll,R.F., *Principles of Interactive Computer Graphics,* 2nd edition, McGraw-Hill, 1979.

[12] Aho,H.V., Hopcroft,J.E., and Ullman,J.D., *The Design and Analysis of Computer Algorithms,* Addison-Wesley, 1974.

[13] Bear,J.L., *Computer Systems Architecture,* Pitman, 1980.

Experience with Specification and Verification
of Hardware Using PROLOG

Norihisa Suzuki
Information Engineering Course
Graduate School, University of Tokyo
Hongo, Bunkyo-ku, Tokyo 113

ABSTRACT

A most important step in VLSI chip design is to write functional specifications very early in the design to debug the top-level protocols. Previously we developed a highly procedural concurrent language Sakura for 'functional specification; we described a complex memory system by Sakura. In this paper we take a completely opposite approach. We used PROLOG, a language based on predicate calculus, to write the behavior of the hardware as well as requirement specifications of the memory system of Dorado, a high-performance personal computer. Even though the styles of these languages are very different, we used similar methods to verify the specifications. We attached executable specifications. We attached executable requirement specifications in the form of input and output specifications. We compare these experiences.

I. INTRODUCTION

We will be putting very complex computer systems into chips that are only available in a very large scale main-frame computer systems of today. In such VLSI chips sophisticated control structures like pipelines and concurrency that are useful in obtaining high-performance will be used extensively. Even in the conventional computers where debugging is easier, taking out all the bugs is a major problem. It was reported that a very complex memory system contained 50 bugs when the first breadboard model was created [2]. Therefore, a most important issue in VLSI chip design is to take out all the bugs before the mask is produced so that we do not have to repeat the expensive process of the mask design. A most important step in creating a correct design is to generate functional specifications in a very high-level language and to debug these specifications against many test data. We developed a VLSI specification language called Sakura based on a highly-typed, algorithmic language Mesa [5]. Detailed description of Sakura was reported previously [7]. In this paper we will take a completely opposite approach to functional specification and verification; we use a much more high-level, functional language PROLOG, which is based on predicate calculus. We specify a complex memory system of a high-performance personal computer Dorado [2]. We will discuss the comparison of two experiences.

I had an opportunity to write microcode for an undebugged, complex pipeline computer [4]. When we encountered the most difficult bug, it took two of us three days to create a test case so that the bug was isolated and regenerated repetitively. The bug only appeared under the following circumstances: the instruction pipeline was first full, then it was emptied by a jump, and at the same time an interrupt occurs. We first noticed this bug only after running a most complicated software system written for the computer; all the other programs including a compiler, an editor, and a loader could run successfully. This experience told us that it is extremely important to have a very large library of software to test new computers in order to make them reliable if we are to use conventional design and debug methodologies. Transistor-level logic simulation is much too slow for complete debugging of VLSI circuit design. Therefore, the only practical approach to create complex, reliable VLSI chips is to create a breadboard model using TTL circuits and to run the software on it. When we are convinced that the breadboard model works reliably, we can create a chip by directly converting TTL circuits to corresponding MOS circuits.

This approach has been working so far but it limits the scope of VLSI chips in three ways:
1) It is a slow and expensive process, since we have to build a TTL prototype.
2) We cannot exploit all the potentialities of MOS circuits; path transistors will not be used. Therefore, the resulting chips can be large and inefficient.
3) Since prototypes are much larger than the chips and cannot be easily replicated, this approach does not work for multiprocessors.
4) If we are to build computers with new architecture, this approach may not usually is very effective, since we do not have a large library of programs.

Alternative approach we are taking is to use functional simulators. We specify the functionality of the computers in a high-level language; test programs are run on this functional specification. Since functional specifications are written in very abstract level they usually run much faster than the transistor-level logic simulators. Thus, we may be able to create VLSI chips without building breadboard prototypes.

Once the top-level protocols are rigorously defined, there are a number of ways to create mask designs that satisfy these protocols. We are confident that DA tools are becoming quite adequate for such tasks. Electric and timing characteristics of building blocks such as adders, shifters, and registers can be precisely obtained by advanced circuit simulation programs on supercomputers. Logical behavior of the entire chip can be rapidly obtained by the logic simulators.

Therefore, a major issue in creating complex VLSI chips is to give the top-level protocol specifications in a very high-level language and verify that the protocol specifications actually satisfy what we want.

We designed a VLSI modelling language called Sakura and specified a complex VLSI multi-microprocessor system [7]. Sakura is a derivative of a strongly typed algorithmic language, Mesa [5]. We added several features that, we felt, are useful for hardware specifications. There was very few real attempt to use a strongly typed language for hardware specification. Many errors, such as misconnections and underspecifications, were found at compile time; we were very satisfied with the use of a strongly typed language.

After the entire specification was written and was successfully compiled, we ran it with several test programs. We discovered a number of bugs in the specification, but most of them were introduced when the formal specification was created from the natural language prose specification.

Another lesson we learned from this exercise was that randomly created test data are not very useful. It is much more effective if we know the implementation and select test data carefully so that critical logic can be tested often.

It is also very important to be able to observe internal states, since the input and output sequences often may not reveal errors. For example, a cache is considered incorrect if it contains two copies of the same main memory block. However, from the outside it is merely a little inefficient cache.

No matter how advanced our simulation becomes, the hardware is just as good as the test data. This is the severe limitation of this methodology. Mos of the widely used computers have hardware bugs; they occur not so often that people are no longer bothered by the bugs. However, if we want to become perfect, the only technique theoretically known to take out all the bugs is verification. Theoretically people think that this is a good idea, and many efforts are spent to make verification practical; so far very few systems really work.

We will explore a methodology in between the functional simulation and the formal verification. We specify hardware just like in the formal verification. We give input and output assertions in predicate calculus. Then, instead of showing that output assertions will be satisfied by hardware for all the inputs that satisfy input assertions, we only show that this relation holds for some selected inputs. We actually run the triple, input assertion, hardware specification, and output assertion, against some test data. The advantage of this method over the functional simulation is that the output data is automatically checked for correctness. The advantage over the formal verification is that, they can be executed so that we do not have to worry about theorem provers.

We chose PROLOG as the language to write input and output assertions as well as hardware specifications. PROLOG is the natural choice for writing executable assertions, since it is based on logic and much research efforts have been done to create efficient and convenient PROLOG programming systems. We chose PROLOG/KR [6], a derivative of PROLOG developed by Hideyuki Nakashima at the University of Tokyo, partly because it is readily available, but also it has LISP as a sublanguage so that inefficient parts can be rewritten in LISP to run faster.

Hardware specification is also written in PROLOG/KR so that input and output assertions can be executed together with the hardware specification.

Even though PROLOG is used in many automatic programming research, the we believe that our way of using PROLOG is new. What we were doing is to execute requirement specifications, which are input and output assertions, together with implementation specifications, which are hardware specifications on test data to check consistency. Because of the characteristics of PROLOG, assertions can be executed directly

We chose Dorado memory system as the target of specification. It was a good choice, since the details of the mechanisms are published, it has a very complex memory system, and is implemented and widely used.

II. BRIEF INTRODUCTION TO PROLOG/KR

PROLOG is a programming language derived from predicate calculus. PROLOG program consists of a set of logical formulas in Horn's form. PROLOG/KR is a derivative of PROLOG developed by Hideyuki Nakashima at the University of Tokyo [6]. The syntax for the basic form is

```
(ASSERT (P *x *y)
        (Q *x *z) (R *z *y)).
```

and means that in order to attain the goal (P *x *y), attain two subgoals (Q *x *z) and (R *z *y). (P *x *y) succeeds if both (Q *x *z) and (R *z *y) succeeds. It is built on top of Utilisp [1], which is an efficient Lisp system developed at the University of Tokyo, and is used at more than 50 computer sites in Japan.

Since PROLOG/KR is embedded in Utilisp, any Lisp functions can be called as PROLOG/KR predicates. Therefore, data structures such as arrays can be used in PROLOG/KR; some inefficient parts can be written in Utilisp to run very quickly.

III. SPECIFICATION METHODOLOGY

The basic strategy for the specification is to describe one circuit or one control stage by one PROLOG Predicate. The input and output terminals of these circuits are parameters of the predicates. Since PROLOG predicates do not have side-effects, they cannot contain internal states. Therefore, we need to use some way of retaining states in order to represent sequential circuits. There are two ways to achieve this in PROLOG/KR; one is to use Lisp functions and arrays, and the other is to pass states or histories of states as parameters. We use both mechanisms.

We consider timing also, so we need to describe when the output will occur according to the input. The timing information is added to histories and passed as parameters.

Consider a simple adder. We can specify it by the following PROLOG/KR statement:

```
(ASSERT (Adder *In1 *In2 *Intime *Out *Outtime)
        (Plus *In1 *In2 *Out)
        (Plus *Intime 5 *Outtime)).
```

(Plus *In1 *In2 *Out) illustrates a way Lisp functions are called. It calls a Lisp (Plus *In1 *In2) and sets the value to *Out if *Out is undefiend previously.

Requirement specifications are the input and output assertions. Input assertion checks the inputs *In1, *In2, and *Intime. It also passes some state information to the output assertion for the later check. We use the parameter *Instate for this purpose.

```
(Pre *In1 *In2 *Intime *Instate)
```

Theoutput assertion is in general of the form:

```
(Post *In1 *In2 *Intime *Instate *Out *Outtime)
```

The test program for adder must execute three predicates in order:

```
(ASSERT (Test *Inl *In2 *Intime *Out *Outtime)
        (Pre *Inl *In2 *Intime *Instate)
        (Adder *Inl *In2 *Intime *Out *Outtime)
        (Post *Inl *In2 *Intime *Instate *Out *Outtime)).
```

IV. DESCRIPTION OF DORADO MEMORY SYSTEM

Dorado is a high-performance personal computer designed and built at the Xerox Palo Alto Research Center [3]. Even though it is a personal computer, it uses the most sophisticated hardware technologies and mechanisms only available in today's large main-frame computers such as ECL logic and cache memory. Unlike most other cache memory systems, Dorado cache retains written data as long as the cache block is not flushed out. Cache keys are virtual addresses so that the virtual address to real address translation takes place only when there is a cache miss.

The memory system is made up by pipeline stages and resources. Pipeline stages provide control and organized into two pipelines: cache pipeline and storage pipeline. The cache pipeline consists of ADDRESS and HITDATA stages and the storage pipeline consists of MAP, WRITETR, STORAGE, READTR1, and READTR2 stages. The organization of these stages are shown in the following figure.

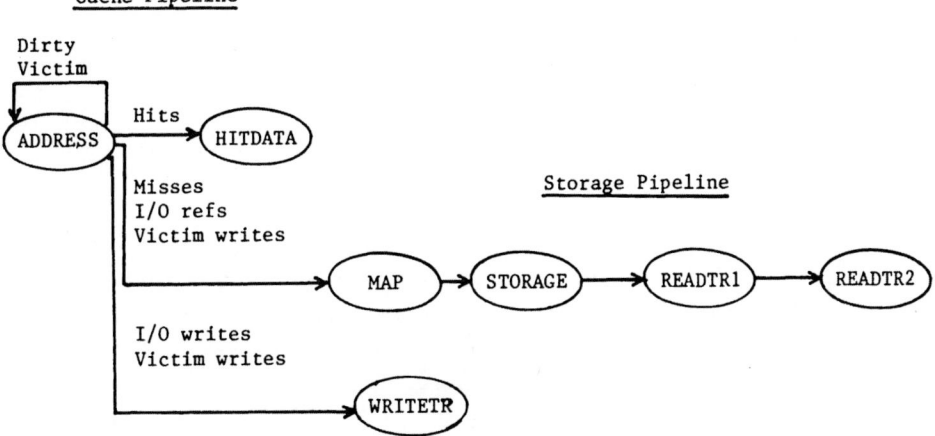

Fig. 1. Organization of Cache and Storage Pipelines

Resources provide the data paths and memories. The major resources mentioned in this section are CacheA, CacheD, StoreReg, FetchReg, MapRAM, WriteReg, StorageRAM, and ReadReg.

4.1. Cache Pipeline

ADDRESS stage
Every memory reference is first handled by the ADDRESS stage. The stage checks whether the virtual address is in the cache; if it is, and the reference is a Fetch or Store, ADDRESS starts HITDATA. ADDRESS starts MAP, if a reference misses or is an I/O reference.

HITDATA stage
The cache address is passed from ADDRESS; the HITDATA stage fetches a word from CacheD storage into FetchReg register if the reference is a Fetch and stores a word into CacheD from StoreReg register if the reference is a Store.

4.2. Storage Pipeline

MAP stage
The MAP stage translates a virtual address into a real address by looking it up in a hardware table called the MapRAM, and then starts the STORAGE stage. MAP takes eight cycles to complete, but starts the STORAGE stage at the fifth cycle.

STORAGE stage
The STORAGE stage is started by MAP: it controls the StorageRAM. STORAGE receives 16 words from WRITETR through WriteReg and sends 16 words to READTR1/2 through ReadReg.

WRITETR stage
The WRITETR stage transports a block into WriteReg, either from CacheD or from an input device; it shares WriteReg with STORAGE. It is started by ADDRESS on every write, and synchronizes with STORAGE. The WRITETR takes at least 11 cycles; the next to the last cycle waits until the third cycle of STORAGE is started.

READTR1 and READTR2 stages
The read operation takes more than 8 cycles because it does error correction and data transport from ReadReg to cache. Therefore, it is split into two stages. On a read READTR1 shifts words out of ReadReg

and through the error corrector. READTR2 reports faults and completes cache read operations either by delivering the requested word into FetchReg (for a Fetch), or by storing the contents of StoreReg into the newly-loaded block in the cache (for a Store).

4.3. Concurrency Control

Since the memory system is pipelined, it is possible that more than one control stages are active. Special mechanisms are implemented to prevent multiple accesses to one resource.

There are three mechanisms to prevent conflicts:

The memory system rejects requests by asserting Hold, if memory system cannot accept a new request.

A reference waits in ADDRESS until its immediate resource requirements are met.

All the remaining conflicts are dealt with in a single state of MAP.

The third stage of MAP, denoted as MAP.3, is the only stage to implement interlocks. The conditions that MAP.3 waits are:

WRITETR activated by the same reference to pass the fifth stage.

If the previous reference is a clean miss and this reference is a clean miss MAP.3 waits two cycles.

When loading and unloading of CacheD occur, the conflicts must be avoided. MAP.3 waits until WRITETR for dirty victim write reference is started.

V. SPECIFICATION OF DORADO MEMORY SYSTEM

We will specify Dorado memory system described in section 4 using PROLOG/KR. In order to make the description simple, we only treated Fetches and Stores in this paper; other types of references can be accommodated very easily. We also made some more simplifications: the cache memory is fully associative and a block of cache contains one word.

5.1. Memory System

The top-level predicate corresponding to the entire memory system is called MemorySystem. The description closely. parallels the organization of the Dorado memory system control structure shown in Figure 1.

```
(ASSERT (MemorySystem *va *command *inData *inTime *outData *outTime
          *fault *inHis *outHis)
    (ADDRESS *va *inTime *cacheResult *cacheAddr *finTime *blockRa
          *victimDirty *inHis NIL *thisHis1)
    (ADD1 *finTime *finTime1)
    (PLUS *finTime 4 *finTime4)
    (SELECT *cacheResult
          (HITS (HITDATA *command *cacheAddr *finTime1 *inData *outData
                  *outTime *fault *inHis *thisHis1 *thisHis5))
          (MISSES (MAP *va *ra *fault *finTime1 *finTim4 *mapTime
                  *inHis *thisHis1 *thisHis2)
              (STORAGE *fault *command *ra *mapTime *storageTime
                  *readReg *writeReg *inHis *thisHis2 *thisHis3)
              (READTR1 *fault *command *ra *storageTime *readTr1Time
                  *readReg *writeReg *inHis *thisHis3 *thisHis4)
              (READTR2 *command *va *fault *readTr1Time *pOutTime
                  *readReg *blockRa *inData *outData *inHis *thisHis4
                  *thisHis5)))
          (IF (= *victimDirty TRUE)
              (AND (AddressForVictim *finTime *thisHis5 *dFinTime)
                  (WRITETR *blockRa *dFinTime *dWFinTime  *dMapGoTime
                      *writReg *inHis *thisHis5 *thisHis6)
                  (MAP *va *ra *fault *dFinTime *dMapGoTime *dMapTime
                      *inHis *thisHis6 *thisHis7)
                  (STORAGE *fault *command *ra *dMapTime *dStorageTime
                      *readReg *writeReg *inHis *thisHis7 *thisHis8)
                  (READTR1 *fault *command *ra *dStorageTime *dReadTr1Time
                      *readReg *writeReg *inHis *thisHis8 *thisHis9)
                  (READTR2 *command *va *fault *dReadTr1Time *dOutTime
                      *readReg *blockRa *inData *outData *inHis *thisHis9
                      *thisHis10)
                  (= *thisHis *thisHis10) (= *outTime *dOutTime))
              (= *thisHis *thisHis5) (= *outTime *pOutTime))
          (Append *inHis ((*command . *thisHis) *outHis))).
```

The predicate MemorySystem takes 9 parameters: *va, the virtual address, and *command, which is the Processor command and takes either Fetch or Store, are supplied by the processor. *inData is the data sent from the Processor; it has the valid value when the *command is Store. *inTime is the time when the MemorySystem is started. *outData receives the value from the MemorySystem when the *command is Fetch. *outTime is the time when the MemorySystem finishes. *fault returns TRUE if there is a page fault and FALSE otherwise.

There is no mechanism for parallelism in PROLOG, yet we have to describe concurrent behavior of hardware. What we did is to create a list of event, which represent memory operations. A pair of a memory operation name and a list of pipeline-stage names with start and finish time is used to represent an event. The last two parameters of

MemorySystem, *inHis and *outHis, are the input history list and the output history list.

MemorySystem first calls ADDRESS; it returns the results in *cacheResult, *cacheAddr, *finTime, *blockRa, *victimDirty, *thisHis1. *cacheResult is either HITS or MISSES. *cacheAddr has the address of the cache block that contains the data of virtual address *va if *cacheResult is HITS. *finTime is the time ADDRESS finishes. *blockRa contains the victim address if the cache reference misses; *victimDirty is TRUE if the victim is dirty and FALSE otherwise. *thisHis1 is the history of this memory reference.

SELECT statement is a control statement of PROLOG/KR and has the following syntax:

```
(SELECT exp
    (exp1 predicate-list1)
    ...
    (expn predicate-listn)).
```

Exp is unified against exp1, ... , expn in order; if the unification succeeds, the corresponding predicate-list is executed and finishes.

If there is a cache hit, HITDATA is executed; the results are returned by *outData, *outTime, *fault which always returns FALSE, and *thisHis5.

If there is a cache miss, MAP, STORAGE, READTR1, and READTR2 are called in order.

Furthermore, if the victim is dirty and has to be written back to memory, and entire set of stages are called: AddressForVictim, WRITETR, MAP, STORAGE, READTR1, and READTR2. Except for the pipeline interlocks, the description of these stages is straightforward.

Pipeline interlocks provide traffic control so that there are no contentions on the resources.

We will describe an example of a pipeline interlock using ADDRESS.

```
(ASSERT (ADDRESS *inTime *cacheResult *cacheAddr *outTime *victimAddr
        *victimDirty *inHis *inThisHis ((ADDRESS *inTime *outTime) .
        *inThisHis))
    (CacheSearch *va *cacheResult *cacheAddr *victimAddr *victimDirty)
    (StAddrTime *inHis *cacheResult *inTime *outTime))
(ASSERT (StAddrTime ? HITS *inTime *inTime))
(ASSERT (StAddrTime *inHis MISSES *inTime *outTime)
    (Last *inHis (*com . *seq))
    (Choose *seq MAP *mapTime)
    (MAX *mapTime *inTime *outTime))
(ASSERT (StAddrTime NIL ? *inTime *outTime))
(ASSERT (Last (*car . *cdr ) *result)
    (Last *cdr *result))
(ASSERT (Last (*car . *cdr) *car))
(ASSERT (Choose (*car . *cdr) *name *fin) (Choose *cdr *name *fin))
(ASSERT (Choose ((*name ? *fin) . ?) *name *fin))
```

The predicate ADDRESS first looks up the cache by CacheSearch, then it computes the time when the ADDRESS stage finishes by StAddrTime. Since HITDATA should be always available, the ADDRESS terminates immediately if there is a hit. Otherwise it has to wait until MAP is ready. So it searches the history *inHis to see when MAP terminated the last time. The time ADDRESS finishes is maximum of MAP termination time and ADDRESS start time.

In these predicate definitions, ? is the don't care symbol which matches anything. History list is of the form

```
((Fetch (ADDRESS 0 0)
        (HITDATA 1 1))
 (Store (ADDRESS 2 2)
        (MAP 3 10)
          . . .
        (READTR2 ... ))
 (Fetch    ... ))).
```

Since the second definition of StAddrTime looks for the last use of MAP, it has to search the list backward. Moreover, the history list has a two-level list structure: the top level is the command level, and the next level is the pipeline-stage level. The predicate Last searches the top level backward, and Choose searches each pipeline-stage list backward for the stage named MAP.

5.2. The Assertions

The predicate MemorySystem is called from a predicate, MemOp. The reason why we put one more level of predicate is that we inserted assertions in MemOp. The definition of MemOp is:

```
(ASSERT (MemOp *command *va *result *time *fault *inHis *outHis)
        (PreAssert *va *location)
        (MemorySystem *va *command *dummy *time *result *dummyTime *fault
            *inHis *outHis)
        (PostAssert *va *location *result *fault))
```

Now, it is clear that each time a memory operation is performed, PreAssert and PostAssert are called to check consistency. PreAssert finds the place of the memory block that contains data with the virtual address *va. The place is returned in *location.

```
(ASSERT (PreAssert *va *location)
        (CacheSearch *va *cacheResult *cacheAddr ? ?)
        (SELECT *cacheResult
            (HITS (= *location Cache))
            (MISSES (MapSearch *va *mapAddr *fault)
                (IF (= *fault FALSE) (= *location Memory)
                    (= *location Disk)))))).
```

The predicate PostAssert assures that the operation is performed correctly. It succeeds if the cache contains the desired block.

```
(ASSERT (PostAssert *va *location *result *fault)
    (IF (= *location Disk) (= *fault TRUE)
        (AND (CacheSearch *va HITS *cacheAddr ? ?)
            (Aget CacheDataVec *cacheAddr *result))).
```

VI. CONCLUSION

A functional description of a complex memory system of a high-performance personal computer was written in PROLOG. The input and the output assertions were also written in PROLOG so that the consistency between implementation specifications, functional specifications, and requirement specifications, input and output assertions are checked by running the program with some data.

Since we showed that we can write functional specifications using a general purpose language, it is not necessary to create a special purpose functional simulation language. It is very important and useful to write requirement specifications, when we are to create rigorous, reliable functional specifications. One can find out logical flaws in writing requirement specifications. What is more important is that we can easily check the consistency between two specifications by executing them so that bugs are discovered at the early stages of design.

Using PROLOG for functional specifications is new. We found several advantages of PROLOG for such purposes. First of all, it is very easy to write and debug PROLOG programs. It is like writing Lisp, but can use pattern matching and backtracking. We wrote Dorado specification in a week; it took us about a month to describe a memory system of similar complexity in Sakura. The majority of time is spent in debugging when we used Sakura, because we had to rely on Mesa debugger, which does not have a good debugging facility for concurrent programs. In future a crucial factor in the cost of VLSI chip development is the amount of time taken for design. The use of PROLOG may well be very important under such circumstances.

The major drawback is that the current implementation of PROLOG/KR is very slow and space consuming. We had to write critical parts in Lisp in order to complete the simulation. Research in creating an optimizing compiler along the line of D. Warren [8] is crucial.

One surprising result was that the lack of explicit parallelism was not a drawback. Even in the language like Sakura, which uses communicating processes for parallelism, we have to print out the history of events in order to check the correctness. Therefore, keeping history as internal data structure will probably not create much overhead.

ACKNOWLEDGEMENT

The author is grateful to Hideyuki Nakshima for letting him use and helping him to understand PROLOG/KR environment.

BIBLIOGRAPHY

[1] Chikayama, T. "Utilisp Manual," Technical Report METR 81-6, Department of Mathematical Engineering and Instrumentation Physics, University of Tokyo, September 1981.

[2] Clark, D. W. et al. "Memory System of a High-Performance Personal Computer," Technical Report CSL-81-1, Xerox Palo Alto Research Center, January 1981.

[3] Lampson, B. W. and Pier, K. A. "A Processor for a High-Performance Personal Computer," Technical Report CSL-81-1, Xerox Palo Alto Research Center, January 1981.

[4] Lampson, B. W. et al. "Instruction Fetch Unit for a High-Performance Personal Computer," Technical Report CSL-81-1, Xerox Palo Alto Research Center, January 1981.

[5] Mitchell, J. G. et al. "Mesa Language Manual," Technical Report CSL-79-3, Xerox Palo Alto Research Center, April 1979.

[6] Nakashima, H. "Prolog/KR User's Manual," Technical Report METR 82-4, Department of Mathematical Engineering and Instrumentation Physics, University of Tokyo, March 1982.

[7] Suzuki, N. and Burstall, R. "Sakura: a VLSI Modelling Language," Proc. Conf. on Advanced Research in VLSI, Artech House, Dedham, Mass., 1981.

[8] Warren, D. H. D. "Implementing PROLOG-Compiling Predicate Logic Programs," DAI Research Report 39-40, University of Edinburgh, May 1977.

VLSI VERIFICATION and CORRECTION

by J. Paul Roth

IBM Thomas J. Watson Research Center, Yorktown Heights NY 10598

In a computer it is indeed necessary to diagnose *failures,* malfunctions in the hardware. It is also necessary to detect *and repair* discrepancies of the design from the specification, the intended function. Errors in design are made, as is well known, both in hardware and software. The effort to compute adequate testing for failures continues, especially with the advent of VLSI. Little effective treatment has been given to the detection *and correction* of errors in design. We consider methods for the detection and correction of errors in logic, timing, and programming errors.

Verification of the design, especially the hardware, *before fabrication* is a *sine qua non* for a VLSI computer. *Verification* means proof that the design, hardware and software, is correct, i.e. performs the required function. It is strategic that the verification of the total design proceed in stages, similar to that of the evolution of the design.

Considerable progress has been made in logic- and timing-verification. On the processor complex, the IBM 3081, it was estimated that 8.5 years of design and fabrication time were saved as a result of systematic application of these procedures, a substantial achievement. The *correction* of errors discovered by these processes was, however, done manually, and thus subject to possible human errors.

In order to determine the correctness of a logic design, it is mandatory to have a reliable means for comparison: this assumes the form either of *another design,* proven e.g. by extensive simulation and usage to be reliable, or a high-level specification of the design, also of proven reliability.

In the case of high-level specification, a program RTRAN has the ability to transform the high-level specification into a faithful logical implementation. Running time for RTRAN is linear with the complexity of the algorithm. Then in parallel, designers perhaps hundreds of them, translated the same specifications into working logic. Then VERIFY or some equivalent algorithm systematically determined whether or not the manual and the automatic designs were equivalent, hence whether or not the logic design is consistent with the specifications. If they were not, then VERIFY would produce counter-examples to their equivalence, perhaps many such. On the other hand, if they were equivalent, then it would also so determine. Running time of VERIFY varied as the square of the complexity of the logics.

If an inconsistency were discovered, the discrepancy would be reported to the designers, who would be obliged to correct the "mistake", not in general an easy task. A "corrected" design would be resubmitted to VERIFY, to validate the correctness of the change, etc., until a correct design were obtained.

A method of automatic correction of logic was devised which seems competitive in quality, certainly for "small" changes, with manual techniques, and offering speed and error-free designs. The method involves running VERIFY "to the end", so that it computes a cover, which for economy could be in factored form, of all primary- and register-inputs for which the primary- or register-outputs disagree; in general not all inputs are specified. Simple logic is then effected, utilizing these covers, to modify the new design so that it agrees with the old reliable. This procedure is effective, certainly for small changes.

TIMING

Given two banks of registers, each gated at different clock times, the problem is to compute all paths, more generally, *chains* though the logic, from one register bank to another, whose "time of flight" is excessive. This delay depends not only on the individual delays of the logic circuits along the "critical paths, chains", but also on the delays in the wiring connecting them. If such critical chains exist, then they must be removed and replaced by faster hardware. The method which we use to handle this problem goes as follows. The first step is to transform the logic into more compact form, utilizing the technique of *dynamic partitioning.* This method obtains effectively an interconnection of PLAs, with their formation adjusted so that it is guaranteed that they not be excessive in size or numbers. The next step is an optimization of each such PLA. An adaptation of the D-algorithm to this PLA optimization process, enhances the size of problem that can be handled. Next step is a multiple-output factorization algorithm, fast, economical and having the important quality of *level control.* This enables the designer to *shrink* the number of levels of logic and thus to reduce effectively the time delay in the hardware. Next is a program called *CONVERT,* which transforms the abstract And-Or-Not logic into a technology of future IBM Systems: this tranformation is simple and effective. The final computation is the *embedding* of the new revised and faster design into the same subspace of the hardware from which the original, time-faulty design came. This is difficult because in general, the faster logic has more circuits. For this purpose we developed a special embedding algorithm, in which the most "critical" circuits were embedded first, both placed and wired, and then the next most critical, etc. The cell calculus (1980) was used in order to enhance the computations.

ERROR DETECTION AND CORRECTION IN ALGORITHMS

For *errors in programs,* no really effective methods exist to detect them, to say nothing of correcting them. An approach will be described for detection and correction.

We think of *algorithms* rather than programs or hardware, because of the enhanced generality and because we shall move interchangeably from one medium to another.

We require our algorithms to be *regular,* which means, among other conditions, that it has no GOTOs and that it is so formulated that they can be translated into a hardware realization, also called regular, which is so formed that it is always *determinate,* (and for which its failures, at least the benign ones, are testable). Again as with hardware, in order to *decide* whether or not a regular algorithm, an R-algorithm, is correct, it is *mandatory* that one has an alternative algorithm, that is alleged to realize the same function reliably. What we have is a "regular" compiler, an R-compiler, which tranforms an R-algorithm into regular logic, an R-design. Call the two R-algorithms a and b, and call their corresponding R-designs, A and B. Such a compiler was written by the combined work of Harry Halliwell and Leon Levy. Our criteria for comparison of A and B and hence a and b are stringent. We assume that the "feedback" variables, whether in the R-algorithm or the R-design, are in one-to-one correspondence, as well as their primary inputs and outputs, and further that, in order for the two R-algorithms, R-designs, to be declared equivalent, we require not only that their corresponding PIs perform the same function but *also* that their feedback, register variables, define identical functions. This eliminates the necessity to perform recursive, sequential computations: it enormously simplifies our task. For this purpose we use VERIFY, or equivalent algorithm, either to determine their equivalence or to obtain for each primary output PI or register output an expression for the differences. We run VERIFY "to the end", i.e. we get a PLA or factored expression for their discrepancy. We implement a *hardware revision* belonging to the algorithm being modified as with the logic-correction scheme described above. Then we enlist an algorithm called PISTAR, which transforms the revised R-design back up to an algorithmic realization, an R-algorithm, which perfectly mates with the revised hardware design. PISTAR uses a modification based upon dynamic partitioning which allows the treatment of large problems. The user may want his verified R-algorithm translated into more conventional language: it is a simple matter to translate a (revised) R-algorithm into a more conventional language such as PL/I.

REFERENCES

1980. Roth, J. Paul, "Computer Logic, Testing, and Verification", *COMPUTER SCIENCE PRESS,* Rockville, Maryland.

1982. Roth, J. Paul and Leon S. Levy, "Equivalence of Hardware and Software", submitted for publication. Cf. IBM Research RC 9716.

1982. Roth, J. Paul, "Dynamic Partitioning", IBM Technical Disclosure Bulletin, to appear. Cf. also IBM Research RC 9515.

1982. Monachino, Michael, "Design Verification System for Large-Scale LSI Designs", IBM Journal of Research and Development, vol. 26, pp. 100-105.

AUTOMATED LOGIC SYNTHESIS

John A. Darringer

IBM T. J. Watson Research Center
Yorktown Heights, New York 10598
USA

ABSTRACT

It is unlikely that we will be able to utilize the full potential
of VLSI without major improvements in designer productivity. One
approach is to design at a higher functional level and to generate
acceptable implementations automatically from such functional
specifications. Previous attempts at automatic logic generation
have usually produced results that were much more expensive than
manual implementation and have relied on exponential 2-level
minimization algorithms which will not scale to VLSI designs. We
are exploring an approach based on local transformations with
nearly linear run times. A system using these ideas has been built
and used to synthesize several gate-array chips with encouraging
results. This system has been extended to remap implementations to
a different technology and to generate alternative PLA and gate
networks for different performance requirements.

INTRODUCTION

Demand for increasingly complex processors appears as fast as advances silicon chip density make their design possible. The result is a growing burden placed on designers. They are asked to design more complex processors and to deal with an growing list of technology constraints to take full advantage of the technology. Thus far designers have been able to meet this challenge, but it is unlikely that they will be able to continue to do so without new tools to improve their productivity.

One approach to increased productivity is automated logic synthesis. Design would be at the functional level and acceptable implementations would be generated automatically. The complexity of this task depends on 1) the level of the functional specification, 2) the nature of the implementation, and 3) the criteria of acceptance. This paper focuses on our exploration of the feasibility of automated logic synthesis for gate-array implementations. It also describes the application of our synthesis system to technology remapping and to the synthesis of custom FET control logic.

Logic synthesis is not a new problem, it has received must attention in the past. Traditional approaches [1] usually rely heavily on 2-level Boolean minimization techniques. These techniques only address an idealized form of the synthesis problem and, more importantly, do not scale with large designs because the algorithms are exponential in the number of inputs or states.

Heuristic techniques have been used. The DDL work at Wisconsin [2,3], APDL at Carnegie-Mellon University [4], and ALERT at IBM [5] all began with behavioral specifications and produced technology-independent implementations at the level of boolean equations. The results were usually more expensive than manual implementations and did not take advantage of the target technology. For example, the ALERT system was validated on an existing design, the IBM 1800, and the implementation produced required 160% more gates than the manual design [6]. There has been an on-going effort at Carnegie-Mellon University developing tools to support high level design [7,8,9,10] In one experiment [11] the CMU-DA system was used to implement the data path portion of a PDP-8/E. It began with a functional description of the machine and produced an implementation in two technologies of the registers, register operators, and their interconnections. This implementation did not include the control logic to sequence the register transfers. When the target technology was TTL series modules the implementation required 30% more modules than the Digital Equipment Corporation implementation. With CMOS standard cells it required 150% more area than the existing Intersil chip. Other smaller experiments have produced standard cell implementations within 20% of manual efforts [12].

LARGE PROCESSOR DEVELOPMENT

The methodology used in developing large processors provides an excellent setting for studying the feasibility of logic synthesis. Synthesis begins with a functional specification. In the design of the IBM 3081 System a formal functional specification of the machine was written in the form of flowcharts [13]. Accompanying the flowcharts are a set of facility declarations that specify the function's input signals, output signals, and memory elements (registers and flip-flops). A synchronous model is assumed and the flowcharts specify the values of the memories and outputs on each tick of an implicit master clock. The flowcharts contain decision boxes, assignment boxes, and multiple simultaneous branches to specify concurrency. Simulation is used to confirm that the flowchart specifies the desired function.

The 3081 System is implemented using a 704 gate TTL gate array. Gate arrays are becoming an increasingly popular method of implementing logic. Their regular layout permits a high degree of automation of physical design (placement and wiring) and allows the user to focus on the logic design. Still the logic designer must find an interconnection of primitive logic blocks (drivers, gates, flip-flops, etc.) that will achieve the specified function. In addition, he must insure that all technology rules (e.g. fan-in and fan-out constraints) are satisfied, that all path lengths are acceptable, that the implementation is testable, that clock signals are correctly distributed, that all external drivers are of the appropriate type, and that the implementation fits on the allocated chips. Finally, he must document his implementation by preparing detailed logic diagrams.

An important problem is to verify that the logic implementation performs the specified function. This is accomplished by a technique called "Boolean Comparison" [14]. This technique insures the equivalence of the two representations and avoids expensive low-level simulation.

LOGIC SYNTHESIS

In this type of environment the problem is to generate acceptable logic diagram implementations directly from flowchart specifications. We have taken an approach [15] that we believe is significantly different from previous efforts. We view synthesis not as a search for an optimal solution, but as a search for a feasible solution satisfying a large number of constraints. Instead of using exponential global algorithms for finding minimum 2-level implementations, we use a sequence of local transformations to accomplish n-level simplification. Also, by operating on the implementation at several levels of abstraction, the transformations can deal with the many varied constraints of design.

The first step is to translate the flowchart specification into a network of interconnected boxes. This is performed is a straightforward manner. RECEIVER boxes are placed on input signals, SENDER boxes are placed on output signals, and all memory elements are translated into boxes of type REGISTER. Operators in the specification appear as boxes in the network. The decision logic and multiple branches are translated into logic that enables the operators. Methods for this type of translation have been described in [2, 4].

Then a small set of local transformations is applied. These transformations look for simple boolean simplifications such as (A

and ¬A), they perform constant propagation, they look for simple
equivalent expressions such as (A and B) and (B and A), and they do
some high level simplification of parity networks and decoders.
Any disconnected boxes or signals are removed in a manner similar
to dead code elimination. Following this, the remaining operators
are replaced by equivalent networks in terms of AND, OR, and NOT
boxes. Those transformations that operate on AND, OR, and NOT are
applied again. The result is a network of boxes of type RECEIVER,
SENDER, REGISTER, AND, OR, and NOT. While the network has been
simplified, it is not in a normal form.

In the next step the AND, OR, and NOT boxes are replaced by
their NAND or NOR implementations depending on whether the target
technology is a TTL gate array or an ECL gate array. These NANDs
or NORs are idealized, however, in that they have no fan-in or
fan-out restrictions.

Now a set of NAND(NOR) transformations are applied to attempt to
reduce the number of boxes of the implementation without increasing
the number of connections. To accomplish this, some
transformations check the fan-out of the various signals involved,
that affect the number of boxes and signals actually removed. The
transformations are applied repeatedly throughout the network.
Some examples of the NAND(NOR) transformations used in our
experiments are shown below; the NOR transformations are identical
except for the operator.

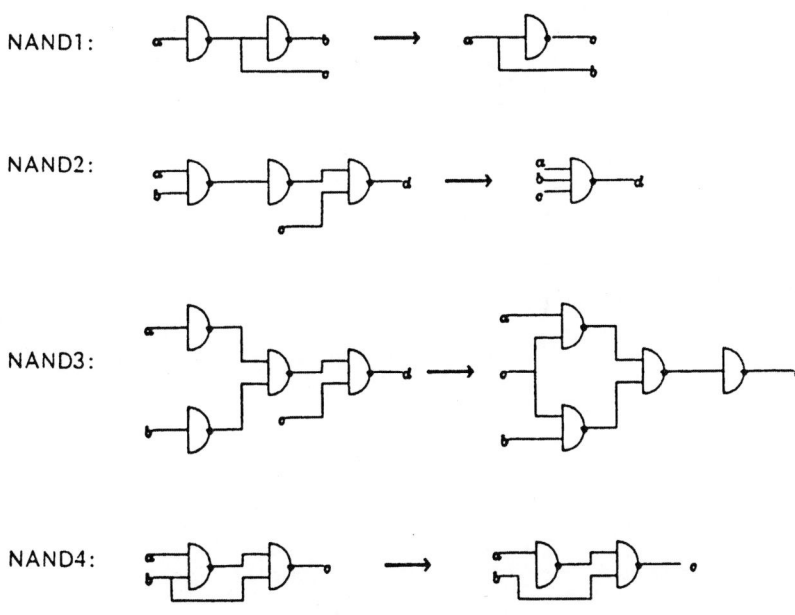

Each transformation has an associated condition that determines if the replacement will simplify the implementation by reducing boxes or connections. These conditions depend on the ·fan-out of the intermediate signals and on whether the target technology is assumed to have dual-rail output. For example, NAND3 is only profitable in certain cases. It does not appear to reduce the box or connection count, but if dual-rail outputs are assumed, the single input NAND on the right hand side is free and disappears after hardware generation. When the NAND(NOR) transformations are complete, a factoring transformation adjusts the fan-in of these boxes to meet the technology constraints, while attempting to share newly created terms. From this point on the transformation become more technology dependent and attempt not to violate previously satisfied constraints such as fan-in.

In the final stage the NAND(NOR) boxes and generic registers are replaced by technology-specific primitives. If the number of control and data lines of the idealized registers exceeds those normally available, additional logic is generated. At this point the implementation is in terms of primitives used by the engineers in their implementations; but because transformations have been made locally, there may be some violations of timing, fan-out, and other technology restrictions. The simplifying transformations at the hardware level are of three sorts. Some are simplifications similar to those at the previous levels, such as eliminating the equivalent of double NOTs, which may occur as a result of expanding higher level boxes. The second type take advantage of the particular technology. For example, flip-flops may provide an output and its complement, allowing some inverters to be removed at this level. Also, a technology may provide primitives combining functions such as flip-flop and receiver that can be used to save boxes. Wired or dotted ANDs or ORs can be introduced to reduce cell count where possible. Some technologies may be dual rail, having both phases available at every gate. This enables simplifications not possible at the technology independent level. The third type of technology-specific transformations used distribute clock signals to flip-flops according to the technology rules, eliminate long and short paths between flip-flops and adjust fan-out by repowering signals.

Several of the transforms at the three levels are analogous, differing only in the types of boxes to which they apply, so that simplifications not made at one level would be caught later. This appears redundant, however, the application of transforms as early as possible helps contain the size of the intermediate results. So, although the same implementation might be produced without the NAND simplifications, they are included for efficiency.

EXPERIMENTAL RESULTS

An experimental system has been built [16]. and has been used to create several chip implementations in two different technologies. In each case, an engineer had implemented the same chip; and we were able to compare the automated design with that of the engineer.

The first experiments with the logic synthesis system were attempts to produce implementations for chips from an existing processor, the IBM 3081, which had been specified functionally and implemented. The target technology was a TTL gate array that provided 96 I/O pins and 704 cells (divided between 3- and 4-input NAND gates) on each chip. In addition to the NAND gates, there are a number of macros such as receivers, senders, and flip-flops that are implemented with these NAND gates. Restrictions on the use of the primitives available, such as fan-in and fan-out requirements, timing constraints, clocking and powering rules were described in a "technology file" and in some cases built into the transformations. Each of the experiments was carried out automatically with the same sequence of transformations, although this particular sequence was the result of much experimentation.

The first chip attempted was a straightforward one. The specification described 7 registers totaling 24 bits, 2 parity operators, and the conditions for the data transfers. The results were that the synthesized implementation was remarkably similar to the manual one. In fact, it required four fewer cells, five fewer connections, and four shorter paths than the engineer's implementation. The similarity, however, was not such a surprise since we had used this example in the design of our system, and since we had worked closely with the chip's designer.

For a second experiment we used the same sequence of transformations on a more complex chip. This chip specification contained 13 register bits, a 3 bit counter, a 5 bit counter, 2 parity operators, and more complex conditions controlling the data transfers. This time there was virtually no contact with the engineer who designed the chip. The synthesis of the second chip resulted in a implementation that required less than 10% more cells and connections than the manual implementation.

To understand how much of our synthesis system was technology dependent, we selected a set of chips that had been implemented is a different technology for our next set of experiments. The target technology was an ECL gate array, which meant that the basic primitive was a NOR and that each primitive had dual-rail outputs; that is, it provides both polarities of its output. The synthesis scenario was adapted to this technology and changed slightly, but the three levels of implementation were maintained. The high level transformations remained unchanged. The NAND level became the NOR level because of the new technology. This required a new transformation to translate the AND/OR primitives into NORs, and a set of NOR simplification transformations. The NAND transformations and the NOR transformations are really the same programs, which operate exclusively on NANDs or exclusively on NORs. The technology-specific transformations had to be rewritten for the new technology, and some new ones were added, such as the one to eliminate inverters.

Experiments were conducted on 5 chips that were all as complex as our first selections. The experiments resulted in implementations with no more that 10% additional gates when compared with their manual implementations.

In all experiments the synthesized implementations were subjected to the same checking that the manual implementations were and no problems were found. The synthesized implementations were functionally equivalent to the specified function; they meet all such as PHILO 18 , offer the designer a choice of PLAs or gates. In this case, the designer is faced with the difficult problem of how to minimize chip area while meeting his performance constraints. It is possible that network of PLAs and gates would solve his problem, but these networks are difficult to design and to analyze.

Daniel Brand has been studying this problem and is exploring an approach based on our logic synthesis system 19 . He begins with a functional specification and transforms it into a network of gates using the technology independent transformations mentioned above. Then two gates are selected that are highly connected, based on a measure of connectivity. An area estimation is made to determine whether the two nodes should be left separate or if they should be merged into a single PLA node. In either case, the result is paired with the next most highly connected node and the process continues until the whole network is visited. The result is a network of PLAs and gates that should have near minimum area. Finally, the program performs a static timing analysis of the network and computes a list of paths that are too long to meet the performance constraints. Then nodes that contain the most critical paths are reexamined and a faster implementation is used if possible. In this way, additional area is used to improve performance. Several experiments have been conducted using these heuristics. In one case, by changing the performance requirements, the program produced six different solutions including 1 PLA with area 3.9 square mils, a network of 12 smaller PLAs with area 1.5 square mils, and a faster network of 3 PLAs with area 2.1 square mils. Other solutions were produced that used combinations of gates and PLAs. While these results are encouraging, the program is very sensitive to estimators of wiring area and PLA size, both of which need further improvement.

SUMMARY

We are in the process of exploring what we believe is a new approach to the old problem of logic synthesis and are encouraged by our initial experiments. We have built a experimental synthesis system and used it to synthesize several gate array chips. In the cases that we were able to compare our results with previous manual implementations, we found that the automatically produced ones required less than 10% additional gates or connections. We have also used our system to remap implemented chips into a different technology, while preserving their input/output behavior. In the area of custom FET implementations, we are looking at methods for generating a network of PLAs and gates to meet given performance constraints with acceptable area. Our hope is that computationally manageable techniques based on local transformations can be used to shorten processor development and validation times.

technology constraints; they meet all design rules; their clock signals were constructed correctly; they were as testable; and while their number of gates and path lengths were sometimes larger, they were all within acceptable limits.

TECHNOLOGY REMAP

With today's rapidly changing technology, a potentially important application for such a logic manipulation system is "remap", which is transforming an existing implementation from one technology into another. A cooperative effort between Mitsubishi and Hiroshima University has developed LORES [17], a system based on macro expansion to assist engineers redesigning existing MSI implementations into LSI technology. Our approach to remapping is not to attempt a one-one mapping of hardware primitives, but to first abstract from the hardware level to the technology-independent NAND or NOR level with generic REGISTERS, DRIVERS, and RECEIVERS. The NANDs (or NORs) can be mapped to NORs (or NANDs) in a straightforward way, and the NAND/NOR and hardware parts of the synthesis scenario can be applied to produce an implementation in the target technology. Only two new transformations were required. One to transform primitives at the hardware level back to the NAND level, and a second to transform the NAND implementation into a NOR one, while preserving the chip input/output behavior. This approach is better than the straightforward replacement of old technology primitives by new ones, because it exposes the remapped implementation to the simplifications at the NOR level and at the hardware level.

To evaluate this approach to remap, experiments were performed to transform 10 chip implementations from a TTL gate array to an ECL gate array. The chips were of comparable capacity and thus chip-to-chip remapping was feasible. Since this chip conversion had not been performed manually we could not make an objective comparison. We did check that the input/output behavior was preserved, that no technology rules were violated, and that the implementation appeared reasonable to an experienced engineer. Chip-to-chip remappings are rare. Usually a new technology will have a different density and number of pins. This could require a merging of several chips from the initial implementation and a partitioning of that remapped, larger function into the chips of the target technology.

CUSTOM SYNTHESIS

While the gate arrays currently used are LSI and there are significant differences between gate array implementations and custom implementations, we believe that the approach presented here can be applied to custom implementations and that the algorithms will scale to VLSI designs. One area that we are exploring is the synthesis of custom FET control logic. A common solution used is to put all control logic in one PLA. Even after minimization the result may be too large and therefore slow, and the designer must decide how to break up the large PLA into a set of smaller PLAs. Occasionally, when performance is very important the control logic may be implemented with an array of gates, where there are known techniques for trading gates for performance. Some technologies,

ACKNOWLEDGMENTS

The work described here is the result of collaboration with several people, principally, Leonard Berman, Daniel Brand, John Gerbi, William Joyner, and Louise Trevillyan. Also Albert Kashner, Scott McFarling, Thomas Wanuga, and Janice Whetsel have made valuable contributions to the design and implementation of the experimental synthesis system.

REFERENCES

[1] M. A. Breuer, Ed., Design Automation of Digital Systems, Prentice-Hall, Englewood Cliffs, NJ, 1972.

[2] J. R. Duley, DDL -- "A Digital Design Language," Ph.D. Thesis, University of Wisconsin, Madison, WI, 1968.

[3] J. R. Duley and D. L. Dietmeyer, Translation of a DDL "Digital System Specification to Boolean Equations," IEEE Transactions on Computers C-18, 305-320 (1969).

[4] J. A. Darringer, "The Description, Simulation, and Automatic Implementation of Digital Computer Processors," Ph.D. Thesis Carnegie-Mellon University, Pittsburgh, PA, 1969.

[5] T. D. Friedman and S. C. Yang, "Methods used in an Automatic Logic Design Generator (ALERT)," IEEE Transactions on Computers C-18, pp. 593-614 (1969).

[6] T. D. Friedman and S. C. Yang, "Quality of Designs from an Automatic Logic Generator (ALERT)," Proceedings of the Seventh Design Automation Conference, San Francisco, CA, 1970, pp. 71-89.

[7] M. Barbacci, "Automated Exploration of the Design Space for Register Transfer Systems," Ph.D. Thesis, Carnegie-Mellon University, Pittsburgh, PA, 1973.

[8] D. E. Thomas, "The Design and Analysis of an Automated Design Style Selector", Ph.D. Thesis, Carnegie-Mellon University, Pittsburgh, PA, 1977.

[9] E. A. Snow, "Automation of Module Set Independent Register-Transfer Level Design," Ph.D. Thesis, Carnegie-Mellon University, Pittsburgh, PA, 1978.

[10] L. J. Hafer and A. C. Parker, "Register-Transfer Level Digital Design Automation: The Allocation Process," Proceedings of the Fifteenth Design Automation Conference, Las Vegas, NV, 1978, pp. 213-219.

[11] A. Parker, D. Thomas, D. Siewiorek, M. Barbacci, L. Hafer, G. Leive, and J. Kim, "The CMU Design Automation System -- An Example of Automated Data Path Design," Proceedings of the Sixteenth Design Automation Conference, San Diego, California, 1979, pp. 73-80.

[12] S. Director, A. C. Parker, D. P. Siewiorek, and D. E. Thomas, "A Design Methodology and Computer Aids for Digital VLSI Systems, IEEE Trans. of Circuits and Systems Vol. CAS-28, No. 7, July 1981.

[13] R. N. Gustafson and F. J. Sparacio, "IBM 3081 Processor Unit: Considerations and Design Process, IBM Journal of Research and Development Vol. 26, No. 1, Jan 1982.

[14] G. L. Smith, R. J. Bahnsen, and H. Halliwell, "Boolean Comparison of Hardware and Flowcharts, IBM Journal of Research and Development Vol. 26, No. 1, Jan 1982.

[15] J. A. Darringer and W. H. Joyner, "A New Approach to Logic Synthesis," Proceedings of the Seventeenth Design Automation Conference, Minneapolis, MN, 1980, pp. 543-549.

[16] J. A. Darringer, W. H. Joyner, L. Berman, and L. Trevillyan, "Logic Synthesis Through Local Transformations", IBM Journal of Research and Development Vol. 25, No. 4, July 1981.

[17] C. Tanaka, S. Murai, S. Nakamura, T. Ogihara, M. Terai, and K. Kinoshita, "An Integrated Computer Aided Design System for Gate Array Masterslices: Part 1. Logic Reorganization System LORES-2", Proceedings of the Eighteenth Design Automation Conference, Nashville, Tennessee, 1981, pp.59-65.

[18] R. Donze, J. Sanders, M. Jenkins, and G. Sporzynski, "PHILO - A VLSI Design System, Proceedings of the Eighteenth Design Automation Conference, Nashville, Tennessee, 1981, pp. 163-169.

[19] D. Brand, "PLAs verses Random Logic" , Research Report RC9505, IBM Thomas J. Watson Research Center, Yorktown Heights, NY, 1982.

Chapter 5

Application Oriented VLSI

A NETWORK FOR PARALLEL SEARCHING

Mamoru Maekawa

Department of Information Science,
University of Tokyo
7-3-1 Hongo, Bunkyo-ku, Tokyo, 113 Japan

ABSTRACT

This paper shows a pragmatic approach to highly parallel data base machines focusing on search operations. A system of n processors connected as a superimposed tree will process as many requests in about twice $\log_2 n$ transfer time between processors. Since it is not difficult for a powerful microprocessor to transfer a request in a couple of milliseconds, a system of 1024 processors will have a processing rate of up to 20K search operations per second.

1. PRINCIPLES OF PARALLEL DATA BASE OPERATIONS

Numeric and non-numeric parallel computations should be based on different principles. Parallelism in numeric computation is obtained between computing units through which data flow (Figure 1). This data flow approach is a basis of many architectures such as data flow machines [1] and systolic architectures [2]. Parallel computation on data bases, on the other hand, should be based on an approach that keeps data "unmoved." This is important due to the volume of data involved. This approach can be called a "function trigger approach" or "function flow approach" as an analogy, as shown in Figure 2. A simple example of calendar data in shown is Figure 3.

An important issue in a function trigger approach is how to determine a pattern of function triggering. If a pattern is known in advance by factors such as the topology of the network as exemplified by the calendar data shown in Figure 3, a fixed communication network can be set up to trigger functions efficiently. In data base operations, however, function triggering patterns are often determined by the contents of data base. A typical example is a search. Two mechanisms shown in Figure 4 follow a function trigger approach but differ in the manner of triggering. Clearly, the indexed search method is more efficient in the use of processors.

A well known principle for function selection is that a processor selection can be performed with the minimum number of comparisons by using a tree structure, provided that data are sorted. While this principle is widely used in sequential processing, it does not depend on the execution mode; sequential or parallel.

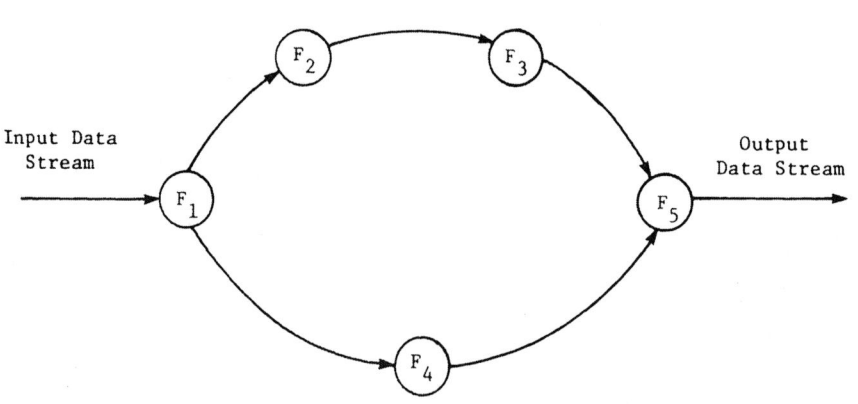

F_i : Operation ; This is activated upon the receipt of its input data

Figure 1.   Data Flow

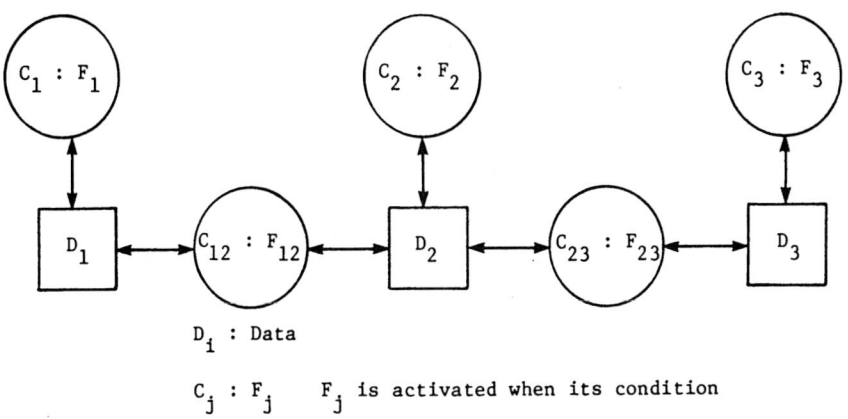

D_i : Data

C_j : F_j F_j is activated when its condition

C_j is satisfied

Figure 2. Function Trigger

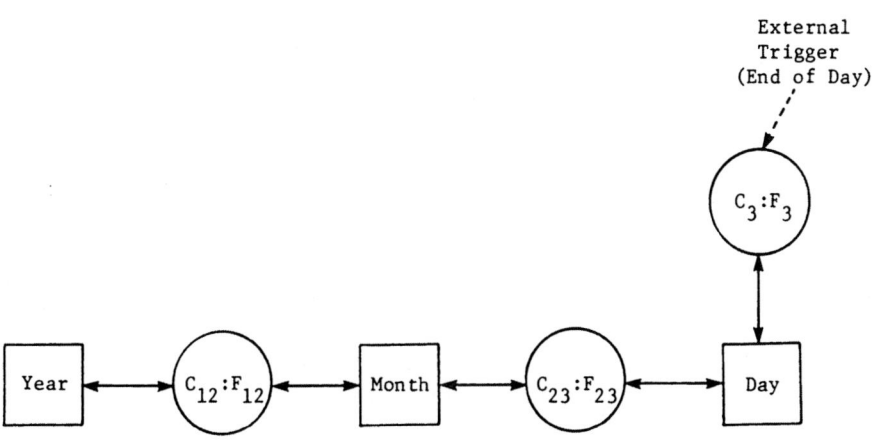

C_3:F_3 End of Day : Increment "Day"

C_{23}:F_{23} Day > Days of Month ("Month") : Increment
                    "Month" & Clear "Day"

C_{12}:F_{12} Month > 12 : Increment "Year" & Clear "Month"

Note : Days of Month *array of* (31, 28, 31, 30, 31, 30, 31, 31,
                           30, 31, 30, 31)

Figure 3. Calendar Data

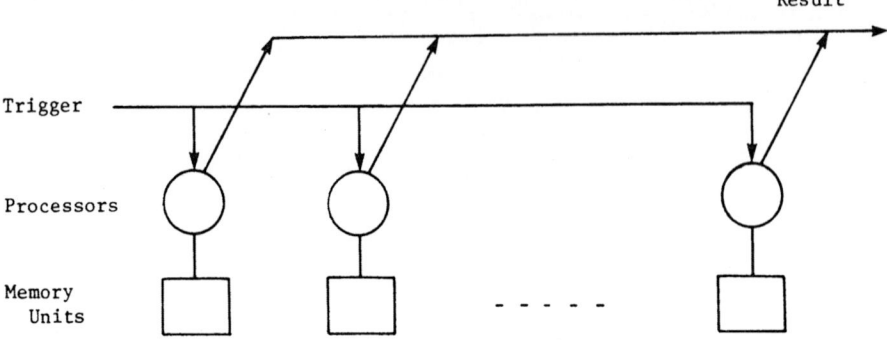

A data base is distributed in the memory units.

(a) Parallel Search

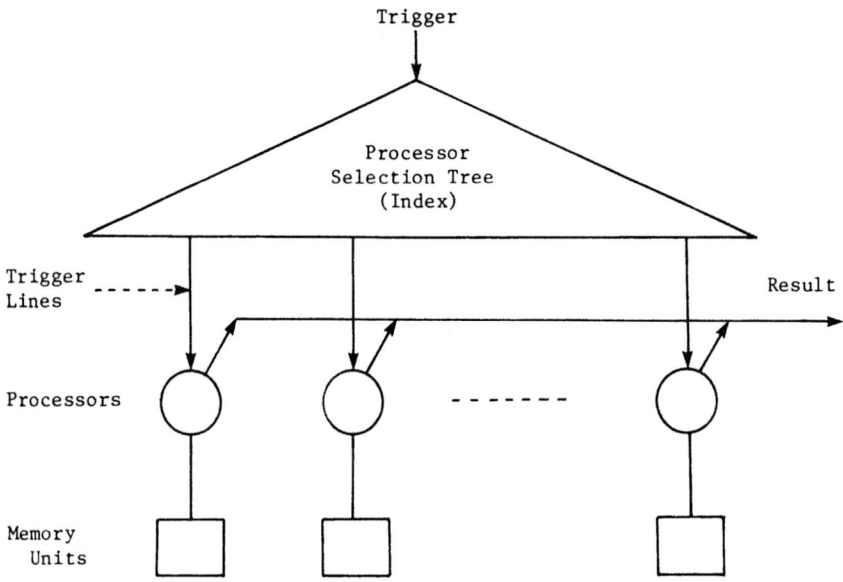

(b) Indexed Search

Figure 4. Two Triggering Methods

2. COMPARISONS

In order to clarify what is stated in section 1, let us analyze a search operation on a data base. Figure 5 shows typical arrangements based on the two approachs. Arrangements a and b are based on the data flow approach. In arrangement b, two variations are shown. Arrangements c through e are based on the function trigger approach. In arrangement c, records are sorted and equally distributed among memory units. A processor will decide for each search key whether a required record is in its memory unit, in a left branch or in a right branch. Arrangement d is the same as arrangement c except that only a single processor is assigned to each level of the tree because only one processor is active at each level of the tree. Arrangement e is based on a superimposed tree [3] and allows n parallel tree searches for a network of n processors. Figure 6 shows superimposed trees of degree 2 and 3 for 8 processors. A superimposed tree of n processors with degree k is a superimposion of n k-ary trees and has the following characteristics:

a) Any processor can be reached in at most $\log_k n$ traverses.

b) Each processor has k input lines and k output lines.

A binary superimposed tree is equivalent to a shuffle- exchange network [4] in the sense that an algorithm on one connection can be one-to-one transformed to one on the other.

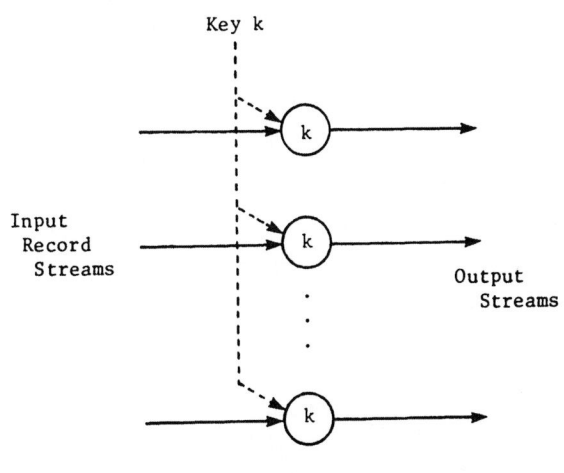

All processors are given the same key.

(a) Partitioned Comparisons

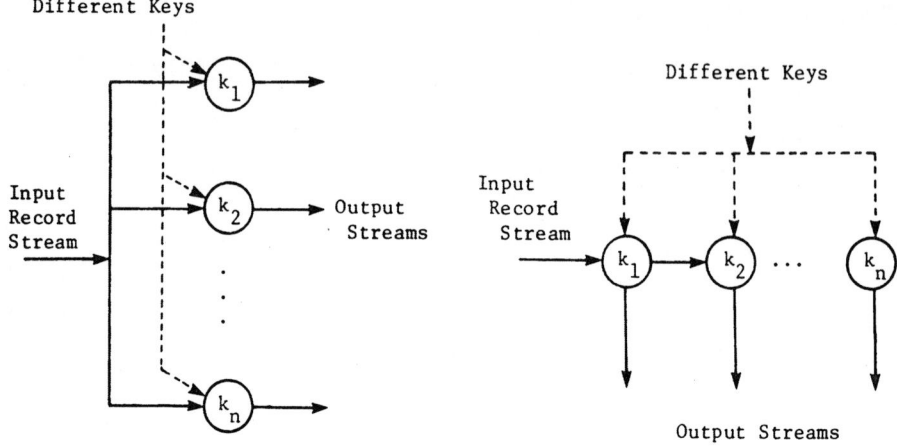

(b) Parallel Comparisons

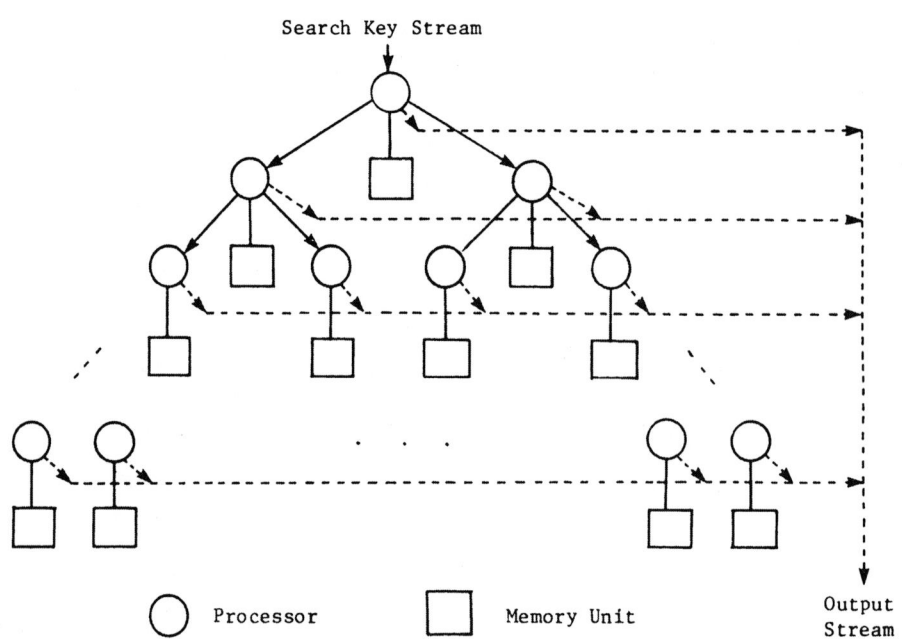

(c) Tree Search

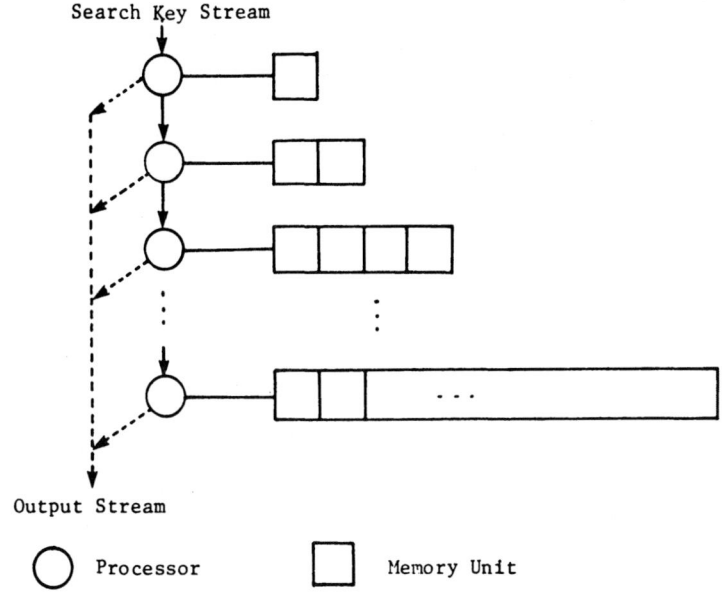

Search Key Stream

⚪ Processor   ⬜ Memory Unit

Output Stream

(d) Pipelined Tree Search

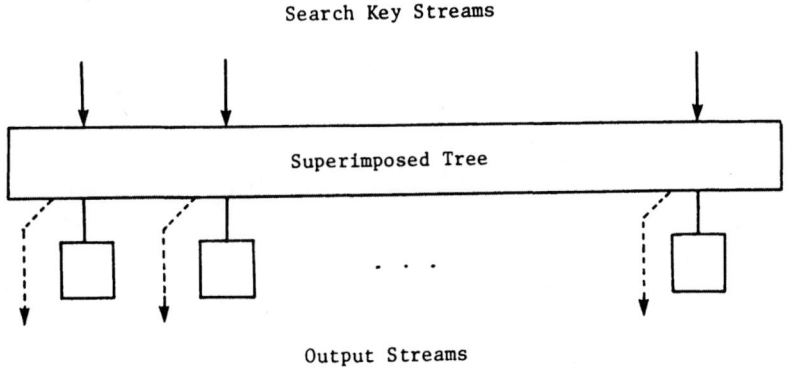

Search Key Streams

Superimposed Tree

Output Streams

(e) Superimposed Tree

Figure 5. Typical Search Methods

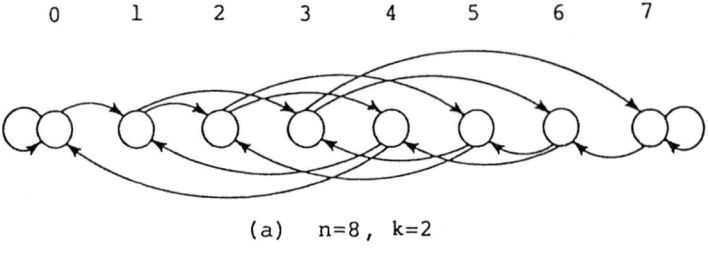

0   1   2   3   4   5   6   7

(a)   n=8, k=2

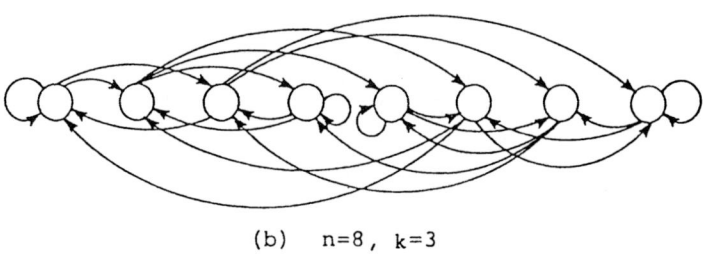

(b)   n=8, k=3

Figure 6.   Superimposed Trees

The characteristics of these arrangements can be defined by the following variables:

n: the number of processing units

m: the number of memory units

r: the number of records

w: record size

P: the time to obtain the result after its request was received

T: the average time to process a request; $1/T$ is the throughput rate

W_I: the input traffic intensity to the network

W_O: the output traffic intensity from the network

U: the internal traffic intensity

$R_n = 1/nT$:the throughput rate per processor

$R_m = 1/mT$:the throughput rate per memory unit

In this definition, small-letter variables are given parameters whereas capital-letter variables represent function values that are determined by computation algorithm and network structure. Table 1 shows the characteristics of the arrangements shown in Figure 5. The following observations can be made:

1) Arrangements a and b, which are based on the data flow approach, have the throughput rates inproportional to the number of records. This is a significant disadvantage.

2) Arrangement c, d and e, which are based on the function trigger approach, have the throughput râtes independent of the number of records. Arrangement d fully utilizes the processors and thus the throughput rate per processor is $1/\log_2 m$. Arrangement e is a parallel version of arrangement d and its throughput rate per processor is $1/h\log_2 m$, where h is about 2 as shown later and is a penalty of parallel processing.

Arrangement	P	T	W_I	W_O	U	R_n	R_m
a $(n=m)$	$r/n+1$	$r/n+1$	nw	hnw $\left(\begin{array}{l}h\text{:rate that a}\\ \text{record matches a key}\end{array}\right)$	$2nw$	$1/(r+n)$	$1/(r+m)$
b $(n=m)$	$r+n$	$r/n+1$	w	w	$nw+1$	$1/(r+n)$	$1/(r+m)$
d	$\log_2 m-1$	1	0	w	$w(\log_2 m+1)$	$1/n$	$1/m$
c $(n=\log_2 m)$	$\log_2 m-1$	1	0	w	$w(\log_2 m+1)$	$1/\log_2 m$	$1/m$
e $(n=m)$	$\log_2 m-1$	$h\log_2 m/m$ (h is about 2)	0	$nw/\log_2 n$	$nw +2nw/\log_2 n$	$1/h\log_2 m$	$1/h\log_2 m$

Table 1. Comparisons of Search Methods

3. SYSTEM STRUCTURE

It is relatively easy to identify basic principles of highly parallel computing on data bases. The next important problem is then to find architectures and algorithms that conform to the restrictions imposed by VLSI technology.

Approaches to VLSI computers can be divided into two groups; one is to design innovative VLSI chips and the other is to construct systems based on powerful but otherwise standard VLSI chips. Our approach is a second approach.

One thing certain about data base processing even in a VLSI era is that data bases must be held in secondary memory such as disks. This remains the case at least in the foreseeable future.

Taking it for granted, our task is to design an efficient index system. A central assumption here is that megabit memory chips are available in quantity so that most crucial indices can be kept in them rather than in disks. Criteria for index system design are physical feasibility and optimality. For the first criterion, we insist that each node of the index system be connected to no more than a fixed number of other nodes. All the structures shown in Figure 5 satisfy this condition. Shuffle-exchange connections [4] and mesh connections also satisfy it.

Optimality criterion requires that the performance/cost ratio be highest for fundamental data base operations. The performance is measured by throughput rate under the condition of unlimited requests whereas the cost is the sum of processor, memory and connection costs. Although there is no agreed set of fundamental data base operations, search and sort operations are unquestionablly two of the most important data base operations. In this paper, we focus on searching because most data base queries are simple searches.

The first design choice for an index system is the number of processors. From Table 1, it is clear that a superimposed tree connection has the highest throughput rate. Its throughput rates per processor and per memory unit are also both optimal. An actual optimal ratio of the numbers of processors and memory units is determined by the relative cast of a processor to a memory unit. Since a processor cost is expected to be smaller than a memory cost, a multi-superimposed tree will provide a higher cost-performance.

Since the number of processors is equal (or proportional) to the number of memory units, a processor (or processors) should be paired with a memory unit so that the function trigger principle is fulfilled and that no complex switching circuit is required. Shared memory structures, for instance, are out of question because of their complex switches.

Summarizing the above discussion, the structure is determined as shown in Figure 7. When it comes to VLSI chip design, it is impossible to contain a superimposed tree in a chip due to pin restrictions. Two possible approaches are:

a)   separate chips for a processor and a memory unit,

b)   a single chip for a pair of processor and memory.

According to the function trigger principle, the second approach is obviously better.

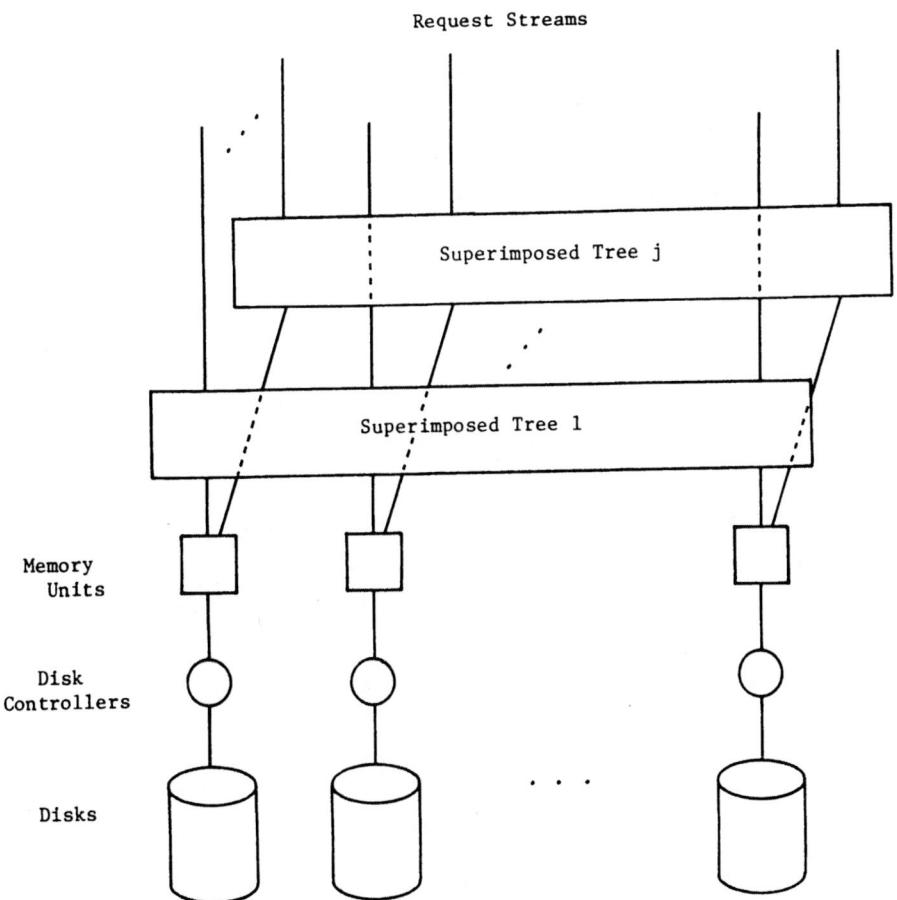

Figure 7.   System Structure

4. ANALYSIS

Algorithm

A search algorithm is a parallel tree search. A case of 8 processors is shown in Figure 8. A search operation of each processor is performed as follows:

(1) Accept a new request from the input line of the processor if the processor has no non-processed request. A non-processed request is a request that has been accepted from the input line yet has not been performed any operation of step 3 and beyond of this algorithm.

(2) Assign $d = \log_2 m$ to the tag of the request. A request consists of identification number, search key, and tag.

(3) Select a request having the minimum tag value among the requests held in this node. If this tag value is 0, the request is sent to the disk controller where requested records are fetched.

(4) Decrement the tag value of the selected request by 1.

(5) Compare the request's key value with the t-th entry of the index of the node where t is the tag value. If the request is greater than or equal to the entry value, the request is sent to node $(2j+1)$ where j is the current node number; otherwise sent to node $2j$.

(6) Repeat steps 3 through 5.

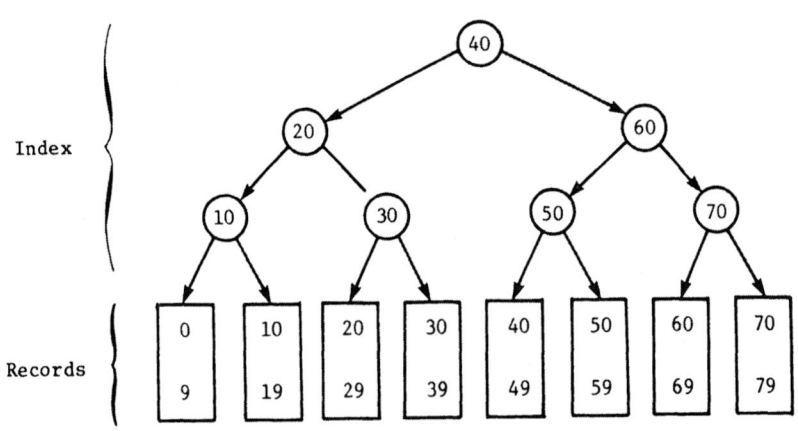

(a) File and Index

Request Streams

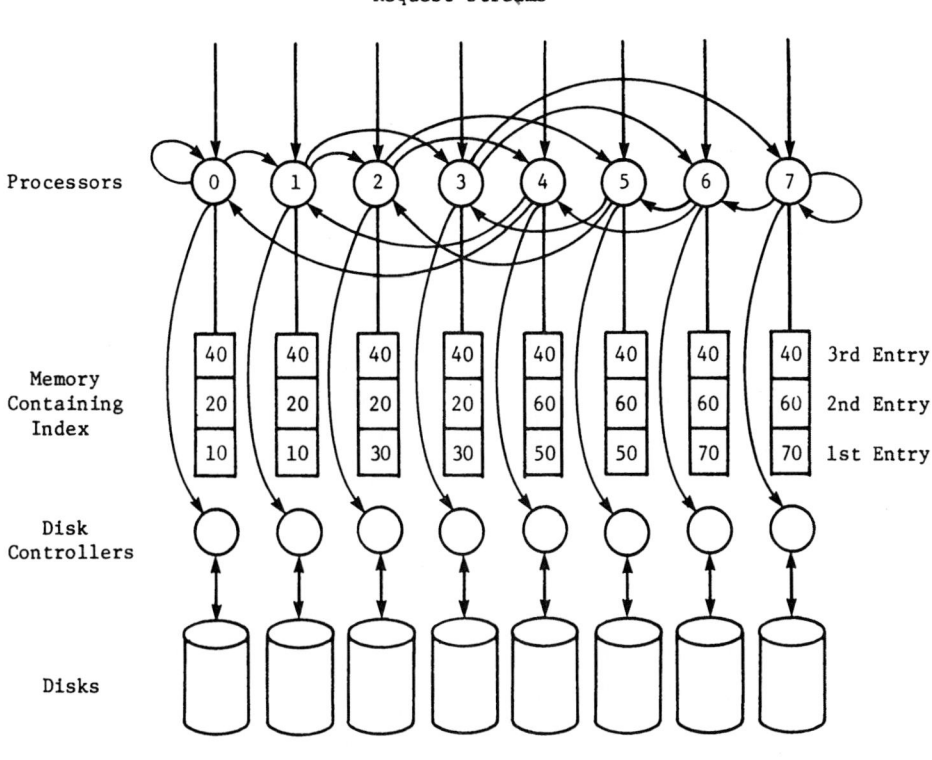

(b) System Structure

Figure 8. Search System

Simulation Analysis

Simulation results of the above algorithm for nT (n times T) are shown in Table 2. Measure nT is the average time to process n requests. The unit of measure is a transfer time of request between nodes and contains steps 3 through 5 of the above algorithm. Steps 1 and 2 are either ignorable or performed simultaneously with other steps.

Table 2 shows that a system of n processors processes n requests in about twice $\log_2 n$ or a little less. Contention to lines and processors reduces the speed to about a half of the theoretical optimal $(\log_2 n - 1)$ but this is still a very favorable result.

n	nT	90% Confidence Interval
64 $(=2^6)$	9.8	±3.0
256 $(=2^8)$	15.1	±3.4
1024 $(=2^{10})$	20.3	±3.8

Table 2. Simulation Results

5. CONCLUDING REMARKS

This paper shows a pragmatic approach to a construction of highly parallel data base machines focusing on searching. It is shown that a system of 1024 processors can process as many requests in 20.3 transfer time. Since it is not difficult for a powerful microprocessor to transfer a request in a couple of milliseconds, 1024 requests can be processed in about 50 milliseconds, which is equivalent to a processing rate of 20K requests per second. A largest problem of constructing such systems is physical feasibility. That is, systems must conform to the practical building restrictions imposed by electronic technology. A construction of such a system is unquestionably a large engineering project.

An even more important issue is whether there exist demands for such powerful data base machines. If there should exist such demands, our approach provides a basic framework for highly feasible and optimal high performance data base machines.

REFERENCES

1. J. B. Dennis and D. P. Misunas, "A Preliminary Architecture for a Basic Data- Flow Processor," Proc. 2nd. Ann. IEEE Symp. Computer Architecture, 1975, pp. 126-132.

2. H. T. Kung, "Why Systolic Architectures," Computer, Vol. 15, No.1 (Jan. 1982), pp. 37-46.

3. M. Maekawa, "Optimal Processor Interconnection Topologies," Proc. Eighth International Symposium on Computer Architecture, Minneapolis, Minnesota, May 12-14, 1981.

4. H. S. Stone, "Parallel Processing with the Perfect Shuffle," IEEE Trans. Computers, Vol. C-20, 1971, pp. 153-161.

A HARDWARE FILE SYSTEM FOR DATABASE STORAGE

Yahiko Kambayashi
Department of Information Science
Kyoto University
Sakyo, Kyoto, Japan

1. Introduction

Due to the recent development of VLSI technology, hardware realization of various functions have been studied by many authors. Such research is especially very important in the areas of databases, picture processing, inference systems etc., where current computer systems do not offer enough efficiency. In this paper a hardware file system for efficient database processing is discussed. One of the most important design criteria of our system is the reduction of communication cost among VLSI circuits, since it is supposed to be a major factor to determine the efficiency of the whole process. First we will discuss VLSI design principles in order to construct a large system using VLSI circuits.

Database machines utilizing associative disk devices and bubble devices have been studied by many researchers [BABB7903][LIN-S7603] [LIPO78][OZKAZ7706][UEMUY8010]. The current tendency is to realize a database machine by VLSI circuits. Circuits for sort, search, select, join etc. are separately discussed[KUNGL8005]. A VLSI database machine is supposed to be realized by a collection of these circuits. Such a realization, however, has the following problem. If there is a circuit to perform a required operation, data must be transmitted to the circuit before the operation and the result has to be transmitted to some circuit or storage. If the operation involves two sets of data, three units of data transmission (two for input and one for output) are required. Another example is as follows. If we have separate sort and search circuits, a new set of data must be transmitted from the sorter to the searcher every time one update operation of data is performed.

Our major design principle is the reduction of the amount of data communication among circuits, which is called <u>the principle of the data transmission cost reduction</u> for VLSI design. It is very important by the following reasons.

A. Since the amount of data to be transmitted in parallel is usually bounded, the transmission time for n data units is O(n).

B. Since the system consists of a lot of VLSI circuits, control of data communication may become very complicated if there is frequent data transmission.

If the transmission time is O(n), utilization of a circuit whose computation time is less than O(n) (O($\sqrt{n}$), O(log n) etc.) may not improve the total performance.

Our approach to meet the principle of the communicatin cost reduction is as follows.

(1) A data storage device should have a function to process data.

(2) Various operations must be realized by a single circuit.

We have designed a functional storage which can realize most of the database operations as well as a data storage function. If a required operation is binary, one set of data is transmitted to the functional storage containing the other set, where the operation is performed. Thus the number of transmission is reduced. We need 2n storage cells to handle two sets whose sizes are at most n.   Since a tree structure is suitable for some database operations and a shift register structure is suitable for some other operations, we use the both structure, each of which has n storage cells. The dual structure makes it possible to realize most database operations efficiently. We expect that by the current tendency of VLSI technology, battery back-uped CMOS circuits are good candidates for nonvolatile storage cells.

If the data to be processed is divided into subsets so that the required operation can be realized by processing each subset separately, each subset can be processed in parallel and thus efficient processing is attained. This is called the principle of partitioning. For the selection operation arbitrary partition is used. For sorting such a partition can be realized by the bucket sort. In our model, there are relations which can not be stored in one functional storage. Such relations are partitioned by key values using the bucket sort. For such purpose we need a hardware index which is a hardware version of an index of a file system. The hardware index can be also realized by the functional storage. Since the system consists of indeces and set of storages, we call the whole system a hardware file system. As the system is realized by VLSI circuits, there are the following advantages over conventional file systems.

(1) Various functions can be realized which cannot be realized by conventional file systems.

(2) Since the index part is realized by hardware, frequent dynamic blancing required by B-trees is not necessary.

There are circuits which require $O(n)$ processing time for any data of size n ($n \leq m$), where m is the maximum size of data processed by the circuit(for example,the systolic array for joins[KUNGL8005]). Since in database systems the data size varies widely, it is better to use a circuit whose processing time is determined by the data size, not by the circuit size.

Other properties required by Iterative Arrays (Hennie) and Systolic Arrays (Kung) are also considered in the system design. There are differences between our design principle and Kung's systolic principle, which is caused by the fact that we emphasize the usage of VLSI circuits in constructing a system and Kung emphasizes the design of a VLSI circuit to perform some specific operation efficiently. For our purpose the reduction of the data communication is very important, but inside of one circuit it may not be so serious. Pipelining operation is very important to improve the efficiency of a circuit but in our case we can use several circuits together instead of using one circuit in pipeline fashion when the latter requires complicate control of data transmission.

In Section 2, definitions on database operations are given. In Section 3, the design principles discussed above are considered in detail. Section 4 discusses the structure of functional storages together with realization of various database operations. In Section 5, the organization and operations of a hardware file system are described.

2. Basic Operations of Relational Databases

A relation R is defined as a finite set of tuples, each of which is a combination of domain values for the attribute set R called a database schema. Fig.1 (a) shows a relation STUDENT. NAME and DEPT are attributes and STUDENT={NAME, DEPT}. There are three tuple in STUDENT. The first tuple (Anderson, Computer Science) shows that Anderson studies at the computer science department.

For a tuple t in R, t[X] denotes the part of t containing only values of attributes in X($X \subseteq R$). The following notations are used for basic relational operations.

Projection: $R[X] = \{t[X] \mid t \in R\}$

Restriction: $R[X \theta C] = \{t \mid t[X] \theta C, t \in R\}$

STUDENT

NAME	DEPT
Anderson	Computer Science
Baker	Physics
Clark	Electronics

(a)

LOCATION

DEPT	BUILDING
Chemistry	B
Computer Science	A
Electronics	A

(b)

R_1

NAME	DEPT	BUILD
Anderson	Computer Science	A
Clark	Electronics	A

(c)

R_2

DEPT	BUILD
Computer Science	A
Electronics	A

(d)

Fig. 1   Examples of relations

θ-Join: $R_1[X_1 \theta X_2]R_2 = \{t_1 t_2 | t_1[X_1]\theta t_2[X_2],$
$t_1 \in R_1, \ t_2 \in R_2\}$

Here, $X \subseteq R$, $X_1 \subseteq R_1$, $X_2 \subseteq R_2$, C is a vector of constants and θ is a comparison operator (=, <, > etc.).

Projection of R on X is obtained by removing all attributes not in X. R_2 in Fig.1 (d) is obtained by a projection from R_1 in Fig.1 (c).

$R_2 = R_1[DEPT, BUILDING]$

The restriction $R[X \theta C]$ shows the subrelation of R consisting of tuples satisfying $X \theta C$. LOCATION and R_2 in Fig.1 (b) and (d) have the following relationship.

$R_2 = LOCATION[BUILDING=A]$

R_1 in Fig.1 (c) is obtained by joining two relations STUDENT and LOCATION in Fig.1 (a) and (b).

$R_1 = STUDENT[DEPT=DEPT] \ LOCATION$

Since the result of the join contains two identical columns, one of them is omitted. Such a join is called a natural join.

For two relations R_1 and R_2 defined on the same attribute set $(R_1 = R_2)$, set operations can be defined. $R_1 \cup R_2$ is a relation consisting of all tuples in R_1 and R_2. $R_1 \cap R_2$ and $R_1 - R_2$ are also defined similarily.

Division is also known as a relational operator which can be expressed by a combination of other operations.

There are aggregate functions such as count, sum and ave. The result of count is the number of different values. For example, $COUNT(LOCATION[BUILDING])=COUNT(\{B,A\})=2$. Sum takes the summation of values and ave calculates average values.

Since contents in a relation can change, we need update

operations such as add, delete and modify. By these operations tuples are added, deleted and modified (a part of a tuple is changed), respectively.

For efficient processing of some of the above operations, we need operations such as sort and search.

These operations are summarized in Table 1, which includes operations not discussed above. In this paper we will discuss VLSI circuits to perform these operations effectively.

```
Basic relational operations
        Projection, Selection, Join, Division.
Set operations
        Union, Intersection, Difference, Direct product.
Aggregate functions
        Count, Sum, Average.
Update operations
        Add, Delete, Modify
Sort and search
        Sort, Direct search, Sequential search.
```

Table 1   Major operations of databases

3. Requirements for Database Hardware

In order to design VLSI circuits for databases, we will discuss properties required for such circuits in this section.

The followings are some of the characteristics of databases which seem to be required to discuss conditions for hardware.

D1 (Data size): The size of data varies from very small to very large.

D2 (Selection): When the data size is large, first the number of data is reduced by selecting tuples satisfying some specific conditions.

D3 (Database operations): Among various operations it is especially important to improve effeciency of sort, search, set operations and joins, which are known as time consuming operations.

Since we have to handle cases when data size is very large(D1), we have the principle of the data transmission reduction(R1) and related requirements(R2 and R3).

R1 (Data transmission cost): If there exists a bound for the amount of data which can be transmitted parallely from one hardware component to the other, the data transmission time is $O(n)$ for data size n. Thus it is very important to reduce the amount of data transmision between VLSI circuits.

R2 (Circuit size): To reduce the necessity of data transmission, a circuit should contain the whole data to be processed. The size of such a circuit is at least $O(n)$, where n is the number of data. Since the difference between computation time of hardware and that of software gets large as n increases, hardware realization is preferable for large n. Thus a circuit whose size is $O(n^p)(p \geq 2)$ does not seem to be practical for replacing software.

For example, since the time required for inputting data to a sorter is $O(n)$, improvement of the processing time over $O(n)$ by sacrificing the circuit size do not seem to be a good method.

R3 (Processing time): If an operation requires input or output of data, it may not be a good method to improve the processing time much shorter than input or output time by using large amount of additional hardware. Reduction of effective processing time by overlapping data input and output processes is very important.

We have the following principle of partition from D2.

R4 (Partition): Selection of data satisfying some specific conditions should be realized efficiently. If the whole data is partitioned into many subsets, selection can be done in parallel at each subset. Most other operations can be realized efficiently by partitioning data.

By D_3, we need circuits to perform database operations. Since we have to reduce the amount of data transmission by R_1, we have the following R_5 and R_6.

R5 (Multiple operations): A circuit which realizes various database operations is preferable. Operations realized by file systems (sort, search, addition and deletion) are very important. Operations should be efficiently performed under condition R3.

R6 (Duplication of storage cells): We need the duplication of the storage cells by the following two reasons. (a) If the operation is binary, the set given from outside should be stored. (b) Since the storage cell should keep a set of data, we need addtional storage to save the result.

When the amount of data gradually increases, there are cases the computation time increases incontinuously. In software realization such a phenomenon occures when the data size exceeds the buffer size for the secondary storage. In hardware realization when the data size exceeds the maximum amount of data handled by a circuit, the time required usually increases very rapidly.

R7 (Expandability): In order to handle data whose size exceeds the maximum ,limit of the circuit capacity, the circuit should be expandable by adding some hardware or there should be a software method to handle the problem efficiently.

Usually in software-based algorithms the computation time is determined by the data size. These are, however, hardware methods whose computation time is determined by the maximum amount of data proceed by the circuit. For example, usually time required by a high-speed 64-bit multiplier is almost fixed even if the inputs are 5-bit numbers. For arithmetic operations, this problem is not serious, since the ratio between the most frequently used data size and the maximum size handled by the circuit is usually not high. As in database system the size of data varies very widely (D1), the problem is serious. When the processing time is $O(m)$ or more (m is the maximum number of data handled by the circuit), the circuit is very much inefficient if the data size n is much smaller than m. The problem is not so serious for circuits requiring $O(\log n)$ or $O(\sqrt{n})$ processing time.

R8 (Data size and processing time): The processing time should be determined by the data size for the process and not by the hardware size.

The following two requirements are conditions for systolic arrays (Kung)[KUNG8201].

R9 (Regularity): A VLSI circuit is required to be realized by a regular arrangement of identical cells. Especially, memory cells and shift register cells can realize higher dencity circuits than other circuits with a similar complexity, it is preferable to use such cells. The system constructed using VLSI circuits should also obey the condition. The system should be built using small variety of different VLSI circuits and connection among them obey some regularity rule.

R10 (Pipeline and parallel processing): To improve the efficiency of a circuit pipeline and parallel processing should be realized.

Pipeling is also very important but not discussed in this paper.

4. Organization of a Functional Storage

A functional storage is desinged to meet most of the requirements discussed in the previous section. From R6 we need 2n storage cells for storing a set consisting of n data. Instead of using two identical n-cell structures, we use an n-cell tree structure and an n-cell shift register structure. Most operations can be realized efficiently by one of the two structures, and during the process the other one can be used to keep the original contents of the functional storage. The proposed circuit satisfies the requirements discussed in the previous section as follows.

(1) R1, R5: The circuit can realize various operations so that data transmission cost will be reduced.

(2) R2, R6: The circuit has a storage function. The number of components is $O(n)$ and the area to embed the circuit is $O(nlogn)$. The original data is kept during the process.

(3) R3, R8: A tree structure is suitable to implement one-tuple search, addition and deletion efficiently. A shift register structure is suitable for sort, sequential search and set operations. The time required to sort is propotional to the data size and not determined by the hardware size.

(4) R9: The system consists of a small number of basic constructs. Especially, most of the system components are functional storages.

R4 and R7 will be discussed in the next section.

Fig.2 (a) shows a basic organization of a functional storage proposed in this paper, for data size n=7. We assume that each cell can store one tuple. If the given relation has more than n tuples, the relation is distributed to more than one functional storage. It consists of a tree part and a shift register part. In order to handle n tuples there are 2n storage cells and $O(n)$ connections, thus the circuit consists of $O(n)$ elements. Since the hight of the tree is $O(\log n)$, the area required for the circuit is $O(nlog n)$, although coefficient part can be reduced by a proper embedding of the circuit (see Fig.2(b)).

We will show how database operations are realized by a functional storage using very simple examples (n=7). In the following discussion, T and S stand for trees and shift registers, respectively, which show the part mainly used by the operation. There are two input/output terminals for the circuit. The terminal for the shift register part is denoted by I/O(S) and the terminal for the tree part is denoted by I/O(T). In order to use the tree part for fast access of data, data should be arranged in ascending or descending order. In the following examples, we use the ascending order only for simplicity.

A. Intial data loading (S): A sequence of data is supplied from
I/O(S).   The sequence starts from L (loading) and end by E (end of
data).   The data sequence can contain B (blank).   Fig.3 shows an
example when 9B7531 is supplied.   The first L sets the operation of
each shift register cell so that only shift operation is realized.
After 6 steps we have the situation shown in Fig.3 (b).   Here E is
supplied from the input terminal, and in this case instead of
shifting data, the contents of the shift register part are copied by
the corresponding tree nodes and L is replaced by B.   The result is
shown in Fig.3 (c).   If the data size is 7, L is shifted out and if
the data size is over 7,   proper warning signal is created.   The
initial loading can be used instead of sort in the following cases.
    a.   Put blanks in the sequence in order to easily handle the
increase of data.
    b. The data are not sorted. For unsorted data usually we cannot use
the tree part for efficient search.   It can be used when the data
are clustered by related key values, etc.
B. Single tuple retrieval (T):If we want to retrieve the data whose
key value is 5 in Fig.3 (c), we put 5 from I/O(T).   This value is
compare with the contents of the root node.   Since 5 is smaller than
7, the left son, the node containing 3 is examined.   As 5 is larger
than 3, the right son is examined, which contains a tuple whose key
value is 5.   Retriving the tuple is done by traversing the path in
the opposite direction (see Fig.3 (d)). The changed data on the path
are recovered by copying data from the corresponding shift register
cells.
C.   Multiple tuple·retrieval (S): If all the tuples are required,
the output terminal for the shift register is used.   After shifting
out (Fig.3 (e)), all shift register cells become empty. The data are
recovered by copying values contained in the tree nodes.
D.   Replacement of blanks (T,S): If a blank node is a parent of a
nonblank node in the tree part,   the search mechanism of the tree
will not work.   Such a tree is called improper.   Improper trees may
be produced by addition or deletion of a tuple.   In a proper tree,
every subtree must satisfy that every node in the subtree is blank
if the root of the subtree is blank. There are two methods to handle
the problem.
    a. Exchange of blank and nonblank values in the tree part.   By a
proper exchange of values,   an improper tree is converted into a
proper tree.   Fig. 4 (c) shows an example of an improper tree.   In
this case exchange of data between the blank node and one of its son

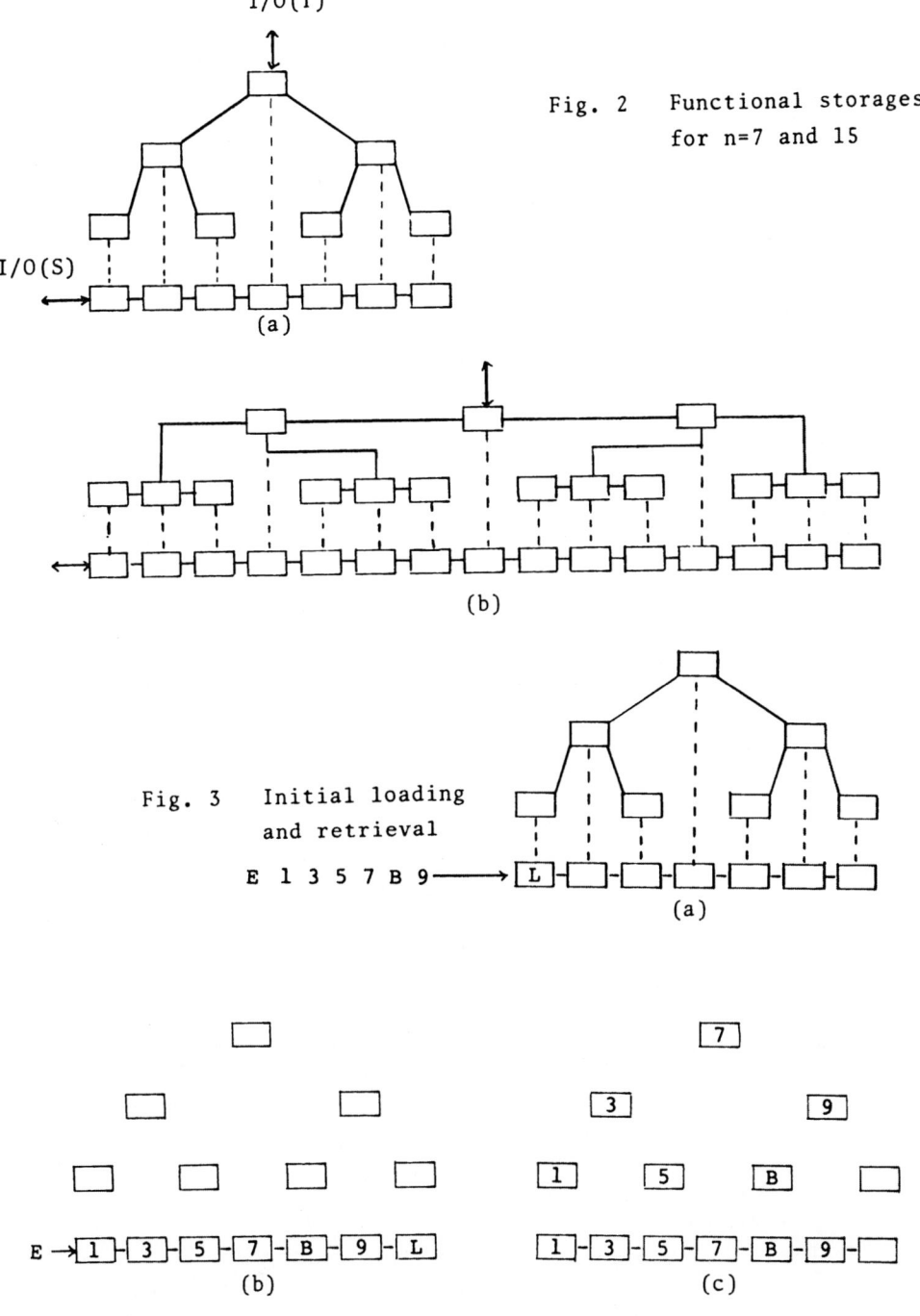

I/O(T)

I/O(S)

(a)

Fig. 2   Functional storages
          for n=7 and 15

(b)

Fig. 3   Initial loading
          and retrieval

E 1 3 5 7 B 9 ⟶ L

(a)

E ⟶ 1 - 3 - 5 - 7 - B - 9 - L

(b)

(c)

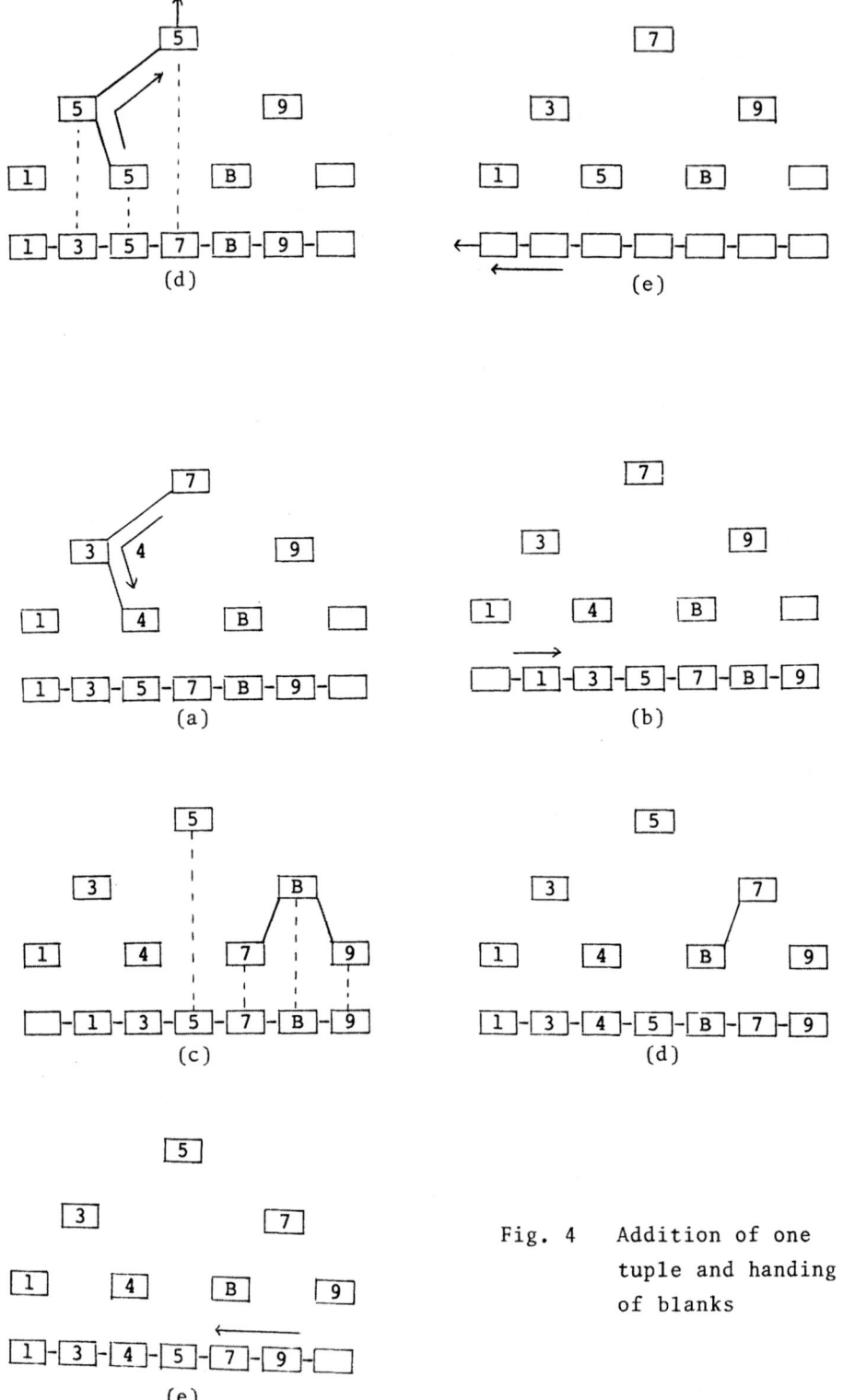

Fig. 4 Addition of one tuple and handing of blanks

generates a proper tree(Fig. 4 (d)).

b. Blank supression (S): The shift register part can be used to remove all blanks contained in the sequence by shifting nonblank values to fill blank values. For example, if the blank supress shift is applied to the circuit shown in Fig.4 (a), the situation shown in Fig.4 (e) is obtained. By copying values in shift register cells blanks in the tree are erased except blanks at the right hand side of the tree.

E. Addition of one tuple (T,S): When it is required to insert a tuple whose key is 8 to the circuit in Fig.3 (c), just apply the same operation as single tuple retrieval and since the left son of 9 is B, the tuple is stored here. When there is no blank cell, we have to use the shift register part. Consider the case when the tuple whose key is 4 is added. Since 4 is smaller than 5, the content of the right son of 3 is replaced by 4 (see Fig.4 (a)). Then the values in the shift register is shifted to the right (Fig.4 (b)), and values in shift register cells are copied by tree node which locate right side of the node 4 (see Fig.4 (c)). In this case replacement of a blank is required(see Fig.4 (d)).

F. Deletion of a tuple(S):Deletion of a tuple is very easily since we only need to replace it by blank symbol B. If the position of blank is not proper one of the two blank replacement operations is applied.

G. Sort (S,T): Sorting is realized by a hardware version of the bubble sort. Tuples are given from the I/O(S) and larger values are shifted to the right. In order to perform sort, the sequence starts from S (Sort). When S passes in the cell, cell opeartion becomes as shown in Fig.5. If the key value stored in a shift register cell in a and the corresponding tree node stores b, the new values for the tree node and the right next shift register cell are c and d, respectively, where

c=min (a,b), d=max (a,b) for ascending order.

S is considered to be larger than blank B and B is considered to be larger than any value.

Fig.6 shows an example when 63714 is an input. Shift register cells are initilized by S which contains (1) the definition of the key and (2) the definition of the ordering, ascending or descending. At Fig.6 (g), S is shifted out. At Fig.6 (k) all the tuples are sorted at the tree part. At Fig.6 (l) values in tree nodes are duplicated to the shift register part and the whole result can be sequentially retrieved from I/O(S). Other operations can be also

215

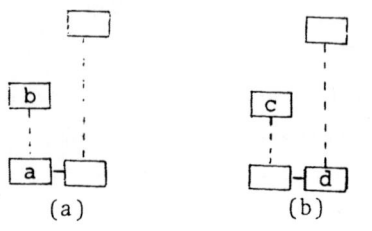

Fig. 5   A basic step
         for sort

(a)   (b)

Fig. 6   An example of
         sorting

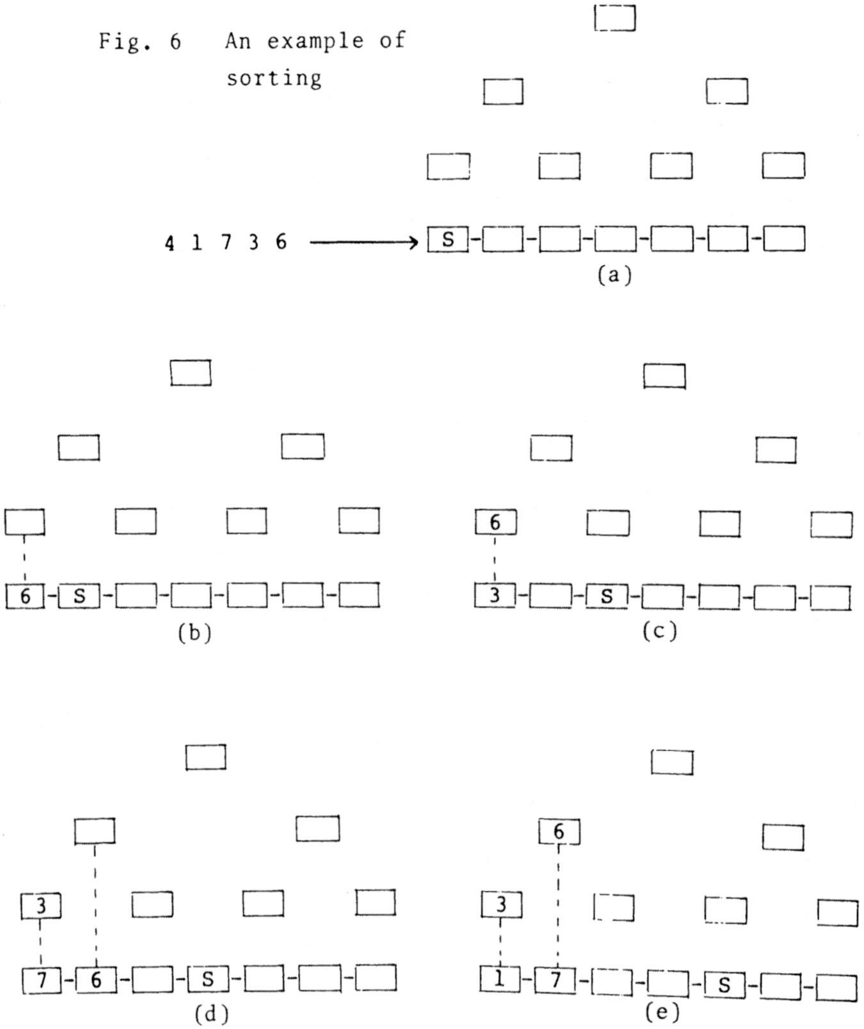

4 1 7 3 6 ──────▶ S-☐-☐-☐-☐-☐-☐

(a)

6-S-☐-☐-☐-☐-☐

(b)

6
3-☐-S-☐-☐-☐-☐

(c)

3
7-6-☐-S-☐-☐-☐

(d)

6
3
1-7-☐-☐-S-☐-☐

(e)

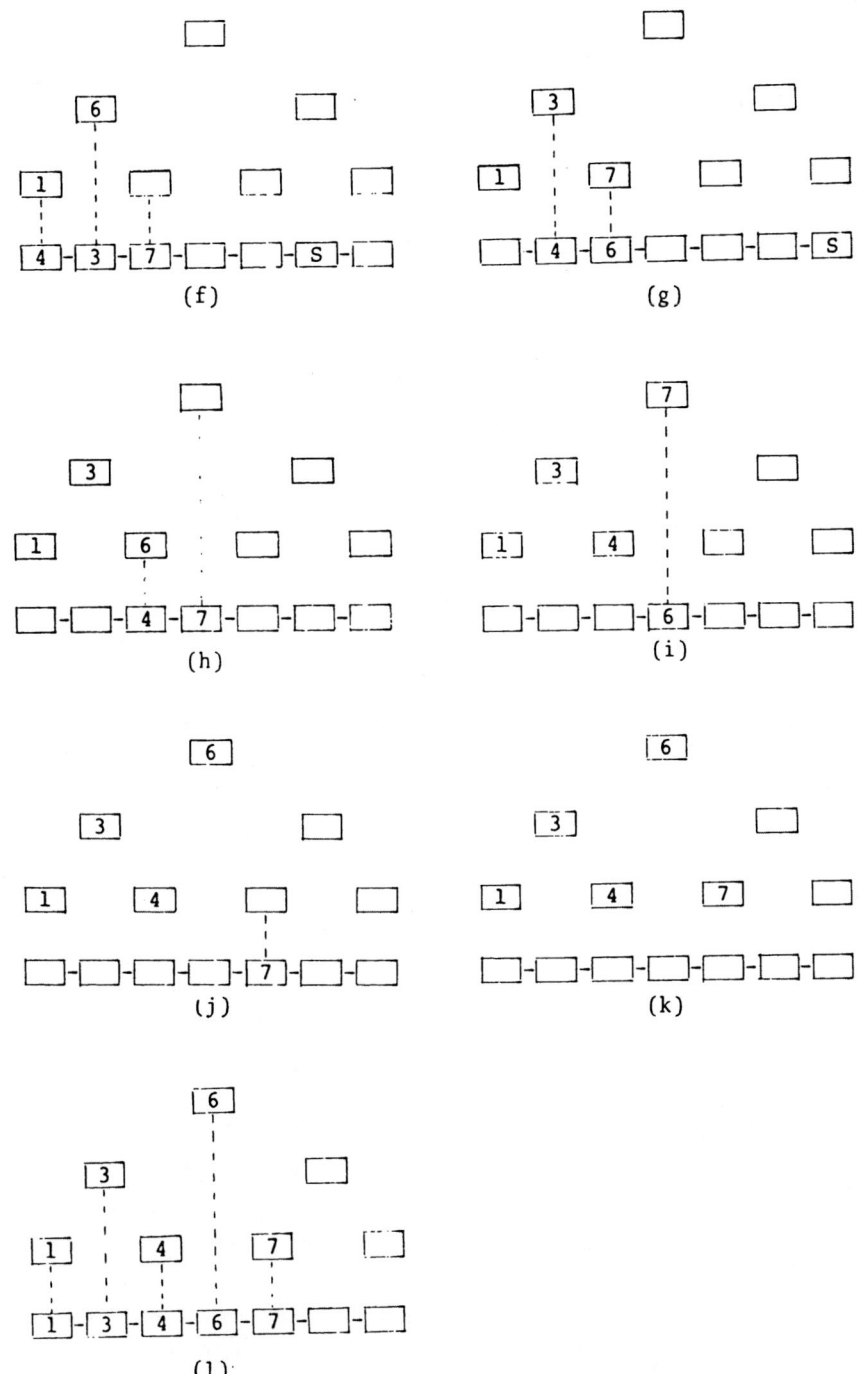

applied to the result.

The circuit can simulate the operation of the bubble up-down sorter developed by Lee, Chang, Wong [LEE-C8106] and and its LSI version by Kikuno, Yoshida, Wakabayashi, Ishikawa [KIKUY8019]. The advantage of the sorter is that as soon as the input is finished we can start to get the sorted result, although the result does not remain in the circuit. Anytime after Fig.6 (f) we can start to get the output. Fig.7 shows the case when the retrieval operation starts from Fig.6 (g). First, values in tree cells and shift register cells are exchanged. Fig.8 shows a basic operation, where c and d satisfy the same condition as Fig.5. Details of the operation is omitted here.

H. Addition of tuples, merge (S,T): By using the sorting function, a set of tuples can be added very easily. This operation can be used for the merge operation for sorted tuples.

I. Deletion of tuples, set subtraction (S): A sequence of tuples to be deleted is given from I/O (S). The top of the sequence is D(delete) to set the cell operation and the last symbol is E. These tuples are shifted to the right and it is examined whether the values at a shift register cell and the corresponding tree cell are equivalent. If these are equivalent, the value in the tree node is replaced by B. The blank removal operation is applied after the deletion. The number tuples contained in the set applied from outside can exceed the storage size n for deletion and intersection, if the set given from outside is not required to be stored.

J. Intersection (S): Intersection is almost same as deletion except tuples replaced by B's in the deletion operation become the results of intersection. Each storage cell contains a tuple and a binary value to indicate the result. The binary values of all cells are initailly 0. Let S_1 be the set stored in the functional storage and S_2 be the set given from the outside. The input is given from I/O(S). The sequence of tuples in S_2 starts from I (to indicate intersection) and ends by E. I contains the information on which part of the tuples to be compared. Tuples in S_2 are shifted to the right and at each step values at each tree cell and the corresponding shift register cells are examined. If these are equivalent, the binary values at both cells are set to 1. After E passes the cell corresponding to the right most nonblank tree cell, the result is obtained as binary values. The binary value for a tuple in $S_1 S_2$ is 1. By the above method we require that n> S_1 + S_2. Another method requiring n>max{ S_1 , S_2 } is as follows. After

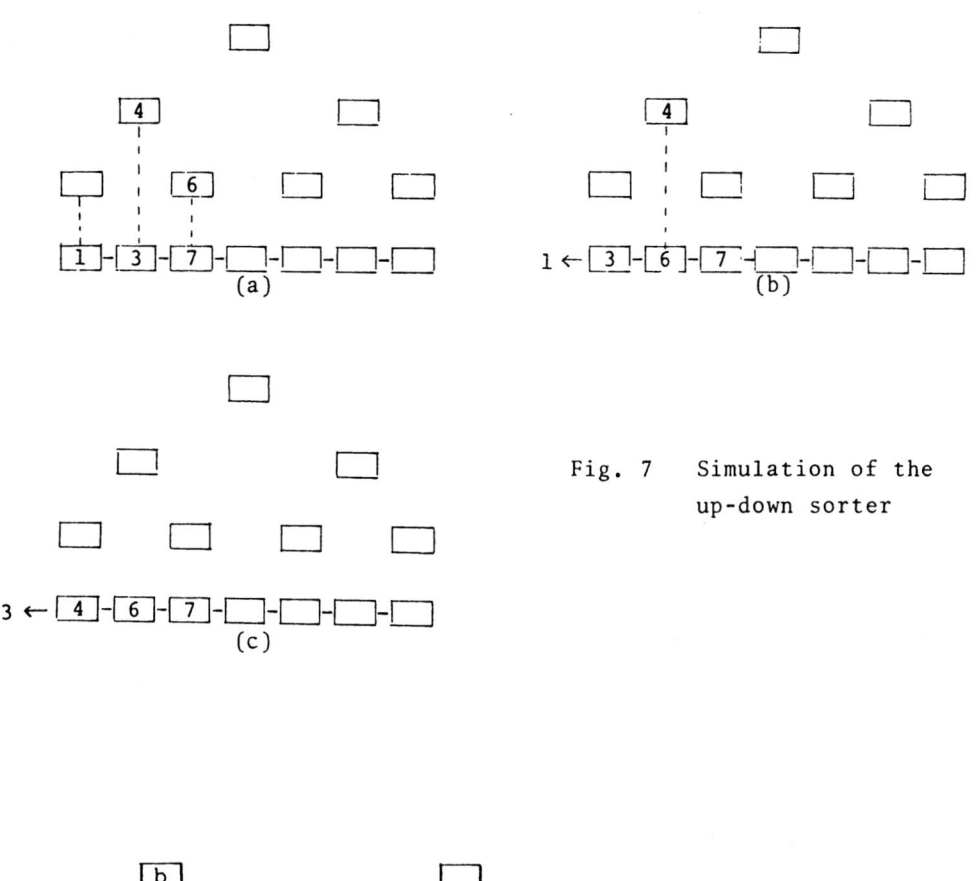

Fig. 7   Simulation of the
up-down sorter

Fig. 8   A basic step for the
up-down sorter

all values in S_2 are given to the functional storage, tree cell values and shift register cell values are exchanged. Then values in shift registers are shifted to the left. At the terminal I/O(S) the binary value is examined and tuples whose binary values are 0 are erased. In this case the result is shifted out from I/O(S). This method can be also used for deletion. In this case tuples in S_2 which are not in $S_1 \cap S_2$ can be detected.

K. Join (S): It is known that any query can be converted into tree queries[SHUM8203][KAMBY8206]. For a tree query there is an effecient procedure for joins utilizing semijoins. Basic operation of a semijoin is intersection of two sets contained in the join attributes. Thus the above intersection procedure can be used for semijoin. We assume that S_1 and S_2 are stored in two different functional storages and the intersection is performed by the functional storage containing S_1. The result is required to be transmitted to the functional storage containing S_2. Since S_2 is stored in the functional storage we only need to transmit the binary values for S_2 which indicate the result. By this way the cost of data transmission can be reduced.

L. Pseudo operations and composite operations (T,S): By using binary values we can indicate tuples satisfying some conditions. More than one conditions can be indicated by permitting more than one binary value for each cell. Operations which do not change the tuple sets are called pseudo operations. Binary values can be also used to realize more than one operations. For example, sorting and intersection can be realized by modifying the sorting operation.

M. Hardware Index: The functional storage can be also used as an index. Fig.9 shows an example. Each tree cell contains two values which show an interval, which are store in the tree to satisfy the ascending order. Each shift register cell contains an address value. For example, if the key value is 5, first the interval store in the root node is checked and then its left son is checked since 5 is smaller than 13. The interval of this node contains 5, so the address is 70. Modification of intervals and addresses can be done by similar methods as discussed above.

There are two modes in the functional storage. The first mode keeps tuples with the same key values and the second mode erases duplicated keys. We can also specify the first key, second key etc. for sorting tuples.

The paragraph before R8 in the previous section shows that (1) if the computation time is O(log n), it is not serious even if n is

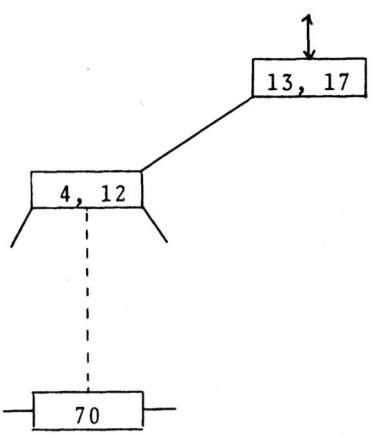

Fig. 9   A hardware index realized by a
functional storage

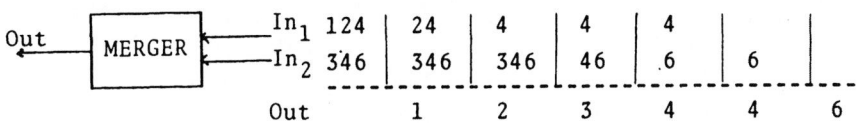

Fig. 10   Merger

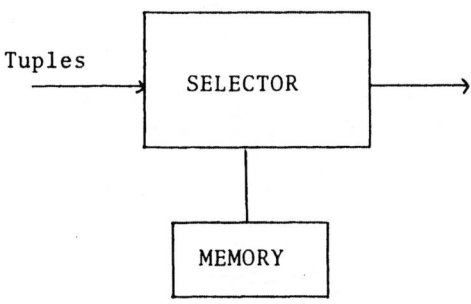

Fig. 11   Selector

the circuit size and (2) if the computation time is O(n), n should be the data size and not the circuit size. The functional storage satisfies these conditions. For operations utilizing the tree part, the computation is O(log m) where m is the circuit size. For operations utilizing the shift register part, the computation time is O(n) where n is the data size.

5. A Hardware File System

A relation with large number of tuples is partitioned into functional storages since the number of tuples to be stored in one functional storage is limited. We will use the bucket sort for the partition.

For each functional storage we determine the upper and lower bounds of key values of tuples. For example, the first functional storage stores tuples whose key values range from A to B and the second functional storage stores tuples whose key value ranges from C to E, etc. The tuple whose key starts from D should be stored in the second functional storage. We assume that a unique order number and a key interval are assigned to each functional storage. These values are stored in the hardware index realized by a functional storage as shown in Fig. 9. The following condition is satisfied by order o(s) and interval i(s) for functional storage s.

For any functional storages s and t,

j < k if o(s) < o(t), j∈i(s), k∈i(t).

Since in each functional storage tuples are sorted, the all tuples in the relation are sorted by retrieving functional storages according to the ascending order of o(s).

A hardware file system consists of a set of functional storages for storing tuples and one hardware index also realized by a functional storage. Compared with conventional file systems, a hardware file system has the following advantages.

(1) Frequent rebalancing operations are not required.

(2) Parallel processing is possible.

(3) Various operations can be realized.

A. Balancing: In B trees dynamic balancing of a tree is required. Since the index is realized by hardware, balancing of the tree is not important. When there are two functional storages with the consecutive o(s) values and the contents of the two can be fit into one functional storage, we can merge them by the merging operation(Section 4, H). If one functional storage becomes full, we only need to prepare another functional storage with the same o(s) and i(s) values. New tuples can be added to either one of them having empty cell space. If a sorted output of the two functional storage is required, we can use the merging hardware to be discussed below (see Fig. 10). Thus we can used more than one functional storage with identical o(s) and i(s) values. If the number of the functional storages with identical o(s) exceeds some predetermined threshold value, we actually need to split these functional storages into storages with consecutive o(s) and i(s) values. Reorganization of the hardware index is not required so frequently as conventional file systems, since usually overflow tuples can be handled without reorganization as discussed above.

B.Parallel processing: The hardware index can be duplicated in order to increase the efficiency. If there are k indeces, the expected sorting time becomes at most k times faster then one index case. Versions of such parallel bucket sorts are discussed by Maekawa et al. [MAEK7911][MAEK8106][WINSC8106][ORENM8110]. Most of other operations can be also done at each functional storage parallely. For example, searching tuples satisfying some condition determined by values of each tuple can be realized at each functional storage independently.

C.Operations: Various functions discussed in the previous section can be realized together with aggregate functions and projections discusséd below. In some cases we can realize a combination of operations by the maximum processing time required by each of these operations. We need additional circuits to realize the whole system.

A merger is used to generate one sorted sequence from a set of sorted subsequences. When these subsequences are given in ascending order, the merger always takes the tuple with the smallest key value among tuples at the top of subsequences. Fig. 10 shows an example of merging of two subsequences. One application of the merger is discussed above. Another application is to sort by values different from the key. When such sorting is required, we first perform sort at each functional storage and then merging of these result is performed.

A selector realizes selection, projection and aggregate functions by processing tuples sequentially(see Fig. 11). Selection is realized by erasing tuples not satisfying the predetermined condition(for example, AGE>20). We can also realize pseudo selection by setting a bit to 0 instead of erasing a tuple. Projection is realized by removing tuple values not corresponding to the given attribute set. Using the memory part of the selector, aggregate functions can be realized. For example, by adding AGE values of the tuples given to the selector, SUM(AGE) can be calculated.

Join of two relations stored in functional storages can be realize by the method discussed by Merrett, Kambayashi and Yasuura[MERRK8109].

We can further generalize the functional storages in order to improve the efficiency by adding (1) bus lines for the shift register part, (2) another shift register and (3) calculation capability to the tree cells.

6. Conclusion

In this paper we showed that (1) in order to realize efficient database systems, component circuits should realize various database functions as well as storage functions and (2) one example of the organization of a such component circuit is shown. Each component circuit should be designed to reduce the data transmission cost among circuits. Currently in the area of VLSI logic design, many circuits realizing single functions are proposed. These results are expected be used to improve the design of functional storages proposed in this paper by a proper combination. By the current tendency of the VLSI technology CMOS is a good candidate for such circuits. CMOS memories are expected to replace dynamic memories for chips over 4M bits and nonvolatite storage can be easily realized by battery back-up, since CMOS consumes very little power.

Another method not discussed in this paper to reduce the data transmission cost among VLSI circuits is utilization of data compression techniques. Data compression methods usually utilize properties of symbols and sequences(for example, see [KAMBN8105]). We can also use properties of operations. For example, the parallel enumeration sorting developed in [YASUT8212] generates a compressed output as a result of sorting. Reduction of data transmission cost for joins utilizing a property of joins is discussed in [KAMBY8206].

The proposed hardware file system has similarity with the current file systems and thus it is considered to be very natural and practical.

Acknowledgement: The author is grateful to Professor Shuzo Yajima, Mr. Hiroto Yasuura and Naofumi Takagi for their discussion. This work is supported in part by the grand from the Ministry of Education, Japan.

References

[BABB7903] Babb, E.,"Implementing a Relational Database by Means of Specialized Hardware," ACM Trans. on Database Systems, vol.4, no.1, pp.1-29, March 1979.

[CHENL7809] Chen, T.C., Lum, V.Y. and Tung, C.,"The Rebound Sorter: An Efficient Sort Engine for Large Files," Proc. Very Large Data Bases, pp.312-318, Sep., 1978.

[CODD7006] Codd, E.F."A Relational Model for Large Shared Data Banks," Comm. ACM, vol.13, no.6, pp.377-387, June 1970.

[DOBO7801] Dobosiewicz, W.,"Sorting by Distributive Partitioning," Information Processing Letters, vol.7, no.1, Jan., 1978.

[ESTEH8005] Estein, R. and Hawthorn, P.,"Design Decision for the Intelligent Database Machine," Proc. AFIPS NCC, pp.237-241, May 1980.

[GOODS8203] Goodman,N. and Shmueli, O., "Transforming Cyclic Schemes into Trees," Proc. ACM PODS, pp.49-54, March 1982.

[HSIA80] Hsiao, D.K.,"Data Base Computers," in Advances in computers, vol.19, pp.1-64, Academic Press, 1980.

[KAMBN8105] Kambayashi, Y., Nakatsu, N. and Yajima, S., "Data Compression Procedures Utilizing the Similarity of Data," Proc. AFIPS NCC, pp.555-562, May 1981.

[KAMBY8206] Kambayashi, Y., Yoshikawa, M and Yajima, S.,"Query Processing for Distributed Databases Using Generalized Semi-Joins," Proc. ACM SIGMOD, pp.151-160, June 1982.

[KIKUY8109] Kikuno, T., Yoshida, N., Wakabayashi, S. and Ishikawa, Y., "A High-Speed Sorting Circuit Suitable to VLSI Implementation," Record of IECEJ SIGAL (Automata and Languages), AL81-50, Sept, 1981. (in Japanese)

[KIM80] Kim, W.,"A New Way to Compute the Product and Join of Relations," Proc. ACM SIGMOD, pp.179-187, 1980.

[KIMK81] Kim, W., Kuck, D.J. and Gajski, D.,"A Bit Serial/Tuple-Parallel Relational Query Processor," Report, 1981.

[KUNG8201] Kung, H.T., "Why Systolic Architecture?," IEEE Computer, vol. 15, no. 1, pp. 37-46, Jan. 1982.

[KUNGL8005] Kung, H.T. and Lehman, P.L., "Systolic (VLSI) Arrays for Relational Database Operations," Proc. ACM SIGMOD, pp. 105-116, May 1980.

[LEE-C8106] Lee, D.T., Chang, H. and Wong, C.K., "An On-Chip Compare/Steer Bubble Sorter," IEEE Trans. Computers, vol. C-30, no. 6, pp. 398-405, June 1981.

[LIN-S7603] Lin, C.S., Smith, D. and Smith, J., "The Design of a Rotating Associative Memory for Relational Database Applications," ACM Trans. on Database Systems, vol. 1, no. 1, pp. 53-65, March 1976.

[LIPO78] Lipovski, G.J., "Architectural Features of CASSM: A Context Addressed Segment Sequential Memory," Proc. Annual Symposium on Computer Architecture, pp. 31-38, 1978.

[MAEK7911] Maekawa, M., "Quick Parallel Join and Sorting Algorithms," Proc. 14th IBM Japan Computer Science Symposium, 1979, Lecture Notes in Computer Science, vol. 133.

[MAEK8106] Maekawa, M., "Parallel Sort and Join for High Speed Database Machine Operations," Proc. AFIPS NCC, vol. 50, June 1981.

[MERR8109] Merrett, T.H., "Practical Hardware for Linear Execution of Relational Database Operations", School of Computer Science, McGill University, Technical Report SOCS-81-30, Sept. 1981.

[MERR8110] Merrett, T.H., "Why Sort-Merge Gives the Best Implementation of the Natural Join", School of Computer Science, McGill University, Techinical Report SOCS-81-37, Oct. 1981.

[MERRK8109] Merrett, T.H., Kambayashi,Y. and Yasuura, H., "Scheduling of Page-Fetches in Join Operations", Proc. Very Large

Data Bases, pp. 488-498, Sept. 1981.

[ORENM8110] Orenstein, J.A. and Merrett, T.H., "Linear Sorting Methods Using Log n Processors", School of Computer Science, McGill University, Technical Report SOCS-81-24, Oct. 1981.

[OZKAS7706] Ozkarahan, E.A., Schuster, S.A. and Sevcik, K.C., "Performance Evaluation of a Relational Associative Processor," ACM Trans. on Database Systems, vol. 2, no. 2, pp. 175-195, June 1977.

[TANAN8010] Tanaka, Y., Nozaka, Y. and Masuyama, A., "Pipeline Searching and Sorting Modules as Components of a Data Flow Database Computer", Proc. IFIP 80, Oct. 1980.

[TANA8208] Tanaka, Y., "Searching and Sorting Hardwares and Their Applications to Symbolic Manipulations," Journal of Information Processing Society of Japan, 23, 8, pp. 742-747, Aug. 1982 (in Japanese).

[THOMK7704] Thompson, C.D. and Kung,. H.T., "Sorting on a Mesh-Connected Parallel Computer", CACM, vol. 20, no. 4, Apr. 1977.

[TODD7809] Todd, S., "Algorithm and Hardware for a Merge Sort Using Multiple Processors," IBM Journal of Research and Development, vol. 22, no. 5, Sep. 1978.

[TONGY8206] Tong, F. and Yao, S.B., "Performance Analysis of Database Join Processors," Proc AFIPS NCC, vol. 51, pp. 627-637, June 1982.

[UEMUY8010] Uemura, T., Yuba, T., Kokubu, A., Ooomote, R. and Sugawara, Y., "Implementation of a Magnetic Bubble Database Machine," Proc. IFIP, pp.433-438, Oct. 1980.

[UEMUM80] Uemura, T. and Maekawa, M., "Database Machine," Published by the Information Processing Society of Japan, 1980 (in Japanese).

[WAH-Y80] Wah, B.W. and Yao, S.B., "DIALOG- A Distributed Processor Organization for Database Machine," Proc. AFIPS NCC, vol. 49, pp. 243-253, 1980.

[WINSC8106] Winslow, L.E. and Chow, Y.C., "Parallel Sorting Machines: Their Speed and Efficiency", Proc. AFIPS NCC, vol. 50, June 1981.

[YAJIY8112] Yajima, S., Yasuura, H. and Kambayashi, Y., "Design of Hardware Algorithms and Related Problems," IECE Japan, SIGAL (Automata and Languages), AL81-86, Dec. 1981 (in Japanese).

[YASUT8212] Yasuura, H., Takagi, N. and Yajima, S., "The Parallel Enumeration Sorting Scheme for VLSI," IEEE Trans. Computers, Dec. 1982(to appear).

Top-down VLSI Design Experiments on a Picture Database Computer

Kazunori Yamaguchi and Tosiyasu L. Kunii

Department of Information Science, Faculty of Science
The University of Tokyo, Hongo, Tokyo 113, JAPAN

Abstract

In this paper, the VLSI implementation of a large system is investigated. Even in the case of VLSI implementation, many chips are required in a large system such as a database system. The interconnection architecture among a large number of VLSI chips and the chip architecture have to be designed to support the logical specification of the system. A sample logical design of such a system is based on our logic PICCOLO (PICture COmputer LOgic) for a picture database computer. If the modularity of our architecture design is good, the interconnection architecture of chips can be decided practically independently of the internal architecture of the chips. In order to support the design, we present a graph based interconnection description diagram and a PASCAL-like programming language. The diagram and language are used to support an extended relational calculus adopted by PICCOLO for describing data operations. An algorithm to decompose the extended relational calculus into the interconnection diagram and the PASCAL-like programming language is also shown. By this decomposition, the performance of the system can be estimated roughly based on an assumption about the properties of the interconnection and the chip before an internal architecture decision.

1. Introduction

Recent technological advancement in VLSI (Very Large Scale Integration) technology has been making a large impact on computer system design methodology. Although the inherent potential of the VLSI technology might free us from the hardware limitations, methodology to design an application oriented system by fully utilizing the VLSI technology has not yet come. The current stage of the VLSI design methodology makes it feasible to design a 16 bit and/or 32 bit microprocessor, a large dynamic/static memory, and fairly complex special purpose chips. The design of the internal architectures of these special purpose chips requires much time if the chip's work cannot be performed by a well-known algorithm such as the Cordic algorithm [AHM82]. Examples of special purpose chips whose internal structures are already well studied are FFT (Fast Fourier Transform) processors for signal processing, voice synthesizers and recognizers, physical and data link layer protocol handlers for local networks, and graphics chips for a fast two/three-dimensional transformation and hidden surface elimination. These chips are designed as general purpose components. In a large system, very many components are required to cooperate with each other. To achieve this goal, the methodology to design functional specifications and internal architectures of chips from a system design is needed.

For the full utilization of the VLSI technology, new methodology named architecture engineering was introduced in our previous paper [YAM81]. In the architecture engineering, we start by deciding on the application oriented hardware architecture. This situation is different from the traditional "grab ready made hardware first and gradually tailor it to the application needs by software" approach. The architecture engineering approach makes a clear-cut contrast to software engineering approach [RAM77] and firmware engineering [AND80]. The architecture engineering we proposed consists of the following four steps.

Step 1. Requirement specification
Requirements for a system are identified.

Step 2. Logical framework decision
Data structures and operations which can support the requirements listed in Step 1 are formalized in well-defined frameworks so that they can be combined consistently. Then, the best framework is determined.

Step 3. Architecture decision
Hardware architectures which can support the framework identified in Step 2 are listed. Then, the best architecture is determined.

Step 4. Design Evaluation
The architecture determined in Step 3 is elaborated and evaluated to show that the resultant design of the system satisfies the requirements.

We take a picture database system as a target system for VLSI design. Up to the present, several picture database systems are constructed to support easy picture handling in geometric data retrieval and medical data handling as shown in Kunii [KUN74, KUN75], Carlson [CAR], Chang [CHA77] and Fu [FU79]. One of the problems of the existing picture database systems is the lack of the appropriate data representation of pictures. Most picture database systems assume that a picture is a two-dimensional array of pixels, and text data and picture data are treated separately. The data operation of such picture databases is a set of picture processing operations decided arbitrarily by the system designer. Most existing picture database systems with formal data models adopt Codd's relational model [COD70]. Because the relational model lacks the capability to represent the structure of the data explicitly, the handling of picture data in the relational model is not an easy task. Another problem of a picture database system is the processing speed. Because the volume of pictorial data is very large, the time required to process picture data is more than that required to process text data by a factor of 10^4 to 10^6. Hence, it is highly desirable to process picture data in parallel to speed up the processing. Although several multiprocessor architectures are proposed for this goal [RIE80, ROE79, SIE79], we actually have two goals to satisfy. One is to support complex picture data structures appropriately. Another is to achieve the high parallelism of data processing. To achieve these two goals at one time, we have designed PICture COmputer LOgic (PICCOLO). We have tested all of the steps of the architecture engineering [YAM80, YAM81] where the last two steps are tested for the implementation case study for the existing parallel processing machine proposed by Guzman [GUZ80]. In this paper, the last two steps are elaborated to show that the picture computer logic PICCOLO is suited for VLSI implementation.

In section 2, the picture database computer PICCOLO we proposed in our previous works is briefly reviewed. Section 3 is devoted to the functional specification of VLSI chips for the picture database computer PICCOLO. Section 4 illustrates the internal architecture of the chips to show that the external functional specifications decided in section 3 can be implemented by simple hardware. Section 5 concludes this paper.

2. PICture COmputer LOgic PICCOLO

2.1. Requirement Specification

First of all, we clarify the requirements for a picture database computer. This is the requirement specification step (Step 1) of the architecture engineering.

For a picture database designed in this paper, a picture may have multiple representations. For example, a picture may consist of a two-dimensional array of pixels, while another picture may consist of objects such as trees and humans, and relationships such as "on the right of" and "below". A picture represented as pixels is an example of a physical picture. A picture represented as objects and relationships is an example of a logical picture. The relationship between logical pictures and physical pictures, and an entire architecture of the picture database, are shown in Figure 1.

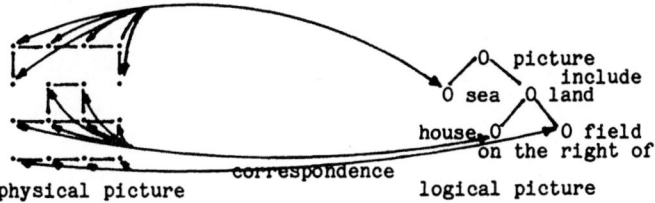

Figure 1. An architecture of a picture database.

As seen in the figure, it is generally the case that a logical picture and a physical picture need to be mutually related somehow. For example, in the case of an image analysis (or pattern recognition) application, it is required to create a logical picture from a physical picture. In case of a computer graphics application, it is required to generate a physical picture from a logical picture. Thus, a physical picture, a logical picture and their relationships need to be handled easily by the picture database.

The requirements for the picture computer are refined into three assertions for the utilization of the VLSI technology as follows:

Requirement 1. The representations of pictures are as uniform as possible.

There are many representations for a picture even if only a physical picture is considered. For example, a two-dimensional array of pixels, a quad tree [ALE78], and an MST (Minimum Spanning Tree) [KRU56, OHB79] are just a few of the popular physical representations. If we can find a formalism or logic to accommodate these varieties of picture representations, only a single hardware architecture (and design) is required for the picture computer. Considering that a mask pattern design for the VLSI occupies the major part of the cost of the VLSI chip, the cost of the VLSI chip is decreased significantly, by reducing the number of hardware designs and hence increasing the production volume for each design.

Requirement 2. Representations and their relationships are treated within the same framework.

By treating a logical picture, a physical picture, and their correspondence within the same framework, we need no additional hardware in between the physical and logical pictures. The relationships among these three are handled by the same hardware which processes the physical and/or logical pictures. Hence, the generation of a logical picture from a physical picture and the opposite are performed within the same framework. It is also possible to derive some new properties of a logical picture such as the distance of objects from the physical picture by utilizing the correspondence between the two.

Requirement 3. The operations are executed concurrently as much as possible.

The number of processors on a single VLSI chip does not necessarily increase its cost significantly. Once the mask pattern design is finished, the chip production cost is essentially independent of the number of processors. Since picture processing generally consists of highly independent operations, such as the update of color values at very many picture points, it is possible to speed up the execution by employing a highly parallel processor chip.

2.2. Definition of PICCOLO

2.2.1. Extension of the Relational Model

As explained in the previous section, the capability to represent an object, a relationship and a relationship among relationships is required for the framework. To represent them, the relational model proposed by Codd [COD70] is extended and named PICCOLO (PICture COmputer LOgic). This is the framework decision step (Step 2) of the architecture engineering.

Definition of PICCOLO

A relation R in PICCOLO is defined as follows:

$$R = N \times N_1 \times R_1 \times N_2 \times R_2 \times \ldots \times N_k \times R_k \times D_1 \times D_2 \times \ldots \times D_m$$

where N, N_i are tuple id domains,

R_i is a set of relation names, and

D_i is a domain.

Suppose that $t = (n, n_1, r_1, n_2, r_2, \ldots, n_k, r_k, d_1, d_2, \ldots, d_m)$ R. Then, n is a tuple id given to the tuple t, n_i and r_i are used as a pair to specify another tuple of a relation r_i with a tuple id n_i, and d_i is a value associated with the tuple t. The tuple t represents a k-ary relationship among tuples which are specified by (n_1, r_1), (n_2, r_2), ..., and (n_k, r_k). From the uniqueness of the tuple id within the same relation, the next condition has to hold.

$$\forall t, t' \in R((n, \ldots) = t \wedge (n', \ldots) = t' \wedge n = n' \rightarrow t = t').$$

We do not impose a stronger condition that asserts the uniqueness of the tuple id in different relations:

$$\forall t \in R, t' \in R'((n, \ldots) = t \wedge (n', \ldots) = t' \wedge n = n' \rightarrow t = t' \wedge R = R'),$$

that is, a tuple id is not unique in the database except within one relation. Hence, it is necessary to use a pair (n_i, r_i) to specify a tuple in the database. By defining tuple ids relative to each relation, the modularity of the system increases.

The values of the tuple id domains are system defined except in case of a generic tuple, and hence **invisible** to the users. The other domains are called **visible**. In this definition, N, N_i are invisible domains, and R_i and D_i are visible domains. To visualize what is represented by a tuple, we illustrate it as Figure 2.

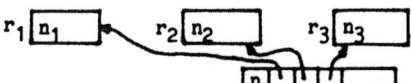

Figure 2. An arrow representation of a tuple $(n, n_1, r_1, n_2, r_2, n_3, r_3, d_1, d_2, \ldots, d_m)$.

In this figure, a tuple $(n, n_1, r_1, n_2, r_2, n_3, r_3, d_1, d_2, \ldots, d_m)$ is illustrated. A pair (n_i, r_i) is represented by an arrow to the specified tuple. We have elaborated this representation and named it an **arrow representation** [KAN82]. A value is not described explicitly if the value is not essential to the explanation.

2.3. Examples of Uniform Representations of Images in PICCOLO

As explained in Section 2.1, varieties of image representation techniques such as the two-dimensional array of pixels, the quad tree and the MST, are in wide use. We show that PICCOLO actually covers all of them. This is the architectural decision step (Step 3) of the architecture engineering.

A. Two Dimensional Array of Pixels

The simplest physical picture is a two-dimensional array of pixels. The array is represented in PICCOLO as in Figure 3.

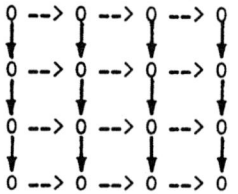

↓ a "below" relationship

-> an "on the right of" relationship

Figure 3. A two dimensional array of pixels represented by objects "O" and relationships "|", "->".

In this description, the pixels are objects and "below" and "on the right of" are relationships among the objects.

B. **Quad Tree**

The quad tree [ALE78] is represented in PICCOLO as in Figure 4.

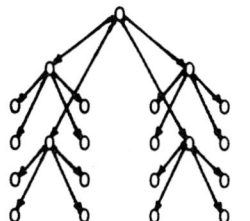

O   a pixel

⟋   a NW relationship

⟍   a SW relationship

⟍   a NE relationship

⟍   a SE relationship

Figure 4. A quad tree represented by PICCOLO.

C. **MST**

A more interesting structure is the MST [OHB78]. The MST is generated by the following operation [YAM81].

For each pair of adjacent pixels (p,p')
     if |gray level of p - gray level of p'| < criteria
     then p is connected to p'.

The MST operation generates, from Figure 3, connected relationships between pixels. The result of this operation is illustrated in Figure 5.

```
O --- O --- O --- O
|                 |
O     O --- O     O
      |     |
O --- O --- O --- O

O --- O --- O --- O
```

O   a pixel

--- a connected relationship

Figure 5. An MST represented by PICCOLO.

The advantages of representing a physical picture in PICCOLO

We showed in the previous section the advantage of treating a physical picture in the same framework with that of a logical picture. If we can also describe the manipulation of a physical picture by formal operations of PICCOLO, the relation between the manipulation language of PICCOLO and the physical picture operation becomes straightforward as well.

2.4. Generic Representation

2.4.1. Basic Definition

In order to provide the PICCOLO system with more flexibility, we introduce a **generic representation of relations**. In order to understand the necessity of the generic representation, let us look at the case where a logical picture of a sheet of cloth in a flat weave woven by interlacing the warp and the woof is described in PICCOLO. By utilizing the pixel ids of Figure 6, we can derive the gray level of the pixel v from the pixel id n by an expression: $v = (n+(n \text{ div } 4)) \text{ mod } 2$.

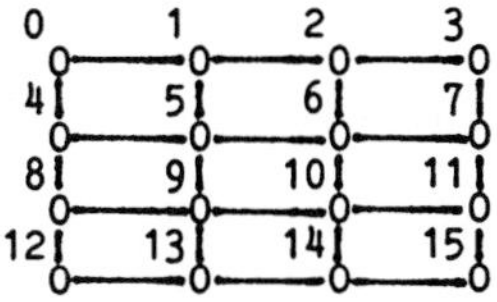

```
0      1      2      3
 o------o------o------o
4|     5|     6|     7|
 o------o------o------o
8|     9|    10|    11|
 o------o------o------o
12|    13|    14|    15|
 o------o------o------o
```

The number marked to the upper left of each circle (pixel) is a pixel id.

Figure 6. A pixel id.

We assume here that the gray level of the warp is one and that of the woof is zero. By utilizing the expression, pixels are represented by one tuple of a relation named a "picture" object as shown in Figure 7.

nid	gray level	COND
n	v	$v=(n+(n \text{ div } 4)) \text{ mod } 2$

A "picture" object

nid1	rname1	nid2	rname2	COND
n	picture	n+4	picture	$n<16-4 \wedge n>=0$

A "below" relationship

nid1	rname1	nid2	rname2	COND
n	picture	n+1	picture	$n<16 \wedge n>=0 \wedge ((n+1) \text{ mod } 4)<>0$

An "on the right of" relationship

Figure 7. Generic representations of relations.

This tuple is called a **generic tuple** and the table is called a **generic table**. In this tuple, n is a variable which runs over pixel ids. COND is a flag to test whether a tuple exists or not for values assigned to the variables. Thanks to variables and COND flags, the single tuple in the "picture" object represents a set of tuples $\{(n, (n+(n \text{ div } 4)) \text{ mod } 2) | 0<=n \wedge n<16\}$. We call the process to obtain a set of tuples $\{(0, 0), (1, 1), (2, 0), ..., (15, 1)\}$ from the generic tuple an evaluation. Similarly, the two relationships "on the right of" and "below" in Figure 3 are represented in Figure 7 as two tuples. One generic tuple of a relation to describe the "below" relationships is equivalent to a set of tuples $\{(n,\text{picture},n+4,\text{picture}) \mid 0<=n \wedge n<16-4\}$, and another generic tuple to describe the "on the right of" relationship is equivalent to a set of tuples $\{(n,\text{picture},n+1,\text{picture}) \mid 0<=n \wedge n<16 \wedge ((n+1) \bmod 4)=0\}$. Hence, a compact representation of relations is possible in PICCOLO after this extension of introducing generic tuples.

2.4.2. Indirect Generic Tuple

We explain the roles of the flag COND in more detail. This flag exists on the domains D_i and N_i. If the value of the COND flag of the D_i is true, then the value in D_i is used as a predicate for obtaining actual values. If the value of the COND flag of the N_i is true, then the value in (N_i, R_i) is used to specify a tuple and the tuple is used as a predicate. This indirection is attractive when there are frequently used predicates. For example, the "below" relation can be represented utilizing indirection as in Figure 8.

```
below  +------+--------+------+--------+-----------------+
       |nid1  |rname1  |nid2  |rname2  |nid3 COND rname3 |
       +------+--------+------+--------+-----------------+
       |  n   |   p    | n+4  |   p    |  m  |condition  |
       +------+--------+------+--------+-----------------+

condition  +-----+------------+
           |nid  |predicate   |
           +-----+------------+
           |  m  |n<16-4 ^ n>=0|
           +-----+------------+
```

Figure 8 An example of indirect COND representation.

2.5. Definition of operations of PICCOLO

In this section, the operations on pictorial data are defined.

Operations to manipulate PICCOLO are based on a relational calculus which are defined below. The syntax of the operations is defined in BNF [HOP79]. The semantics of the syntax are explained in plain English. In the definitions, the notation <a>b is used to stand for a symbol which is of the type <a> and identified by b.

```
<term> ::= <constant> |
           <variable> |
           <function>(<arg>) |
           <term>.<term> |
           <term>◀<term> |
           <term>є_g<term> |

           <quantifier>(<term><term>) |
           (<term>)
<arg> ::= empty |
          <arg><term>
```

An empty and an undef are reserved values.

A <value> is a meaningful unit such as a boolean value, a character string, a binary number, or a relation name, and their meanings are implementation specific. The boolean values, namely, true and false values, are also reserved.

A <variable> is a symbol, and is either free or associated with a value.

A <function> is a function such as an arithmetic operation and its meanings are implementation dependent.

The <term>s <term>.<term>, <term>∈<term>, and <term>∈$_g$<term> are special cases of the <term> "<function>(<arg>)", but because they have database-oriented meanings in PICCOLO, we treat them separately.

<term>t1.<term>t2 gives an attribute value t2 of the relation to which a tuple t1 belongs.

<term>t1∈<term>t2 gives true if and only if a tuple t1 is a tuple of a relation t2, where a tuple of t2 is not evaluated as a generic tuple.

<term>t1∈$_g$<term>t2 gives true if and only if a tuple t1 is a tuple of a relation t2, where a tuple of a relation t2 is evaluated as a generic tuple.

<quantifier>(<term>t1<term>t2) gives a value evaluated from all the values of a term t2 on the condition that a variable-value association satisfies a term t1. The meaning of the <quantifier> is implementation dependent.

3. SPEC: Set Processing Execution Chip

For the purpose of the VLSI implementation of the PICCOLO, we take a top-down design approach. The first step of this design is to design a functional specification of the VLSI chips to support the PICCOLO. The chip which satisfies the functional specification described in this section is named Set Processing Execution Chip (SPEC for short). The internal architecture of the SPEC is designed in the next step.

Even if the amount of hardware on one chip grows large, the system cannot be implemented only on a single chip. One of the reasons for this is that the size of the logic which can be fabricated on one chip is limited even in case of the VLSI implementation. Another is that the number of interface pins of one chip is limited. It is true that the possible number of pins of one chip is increasing. However, the cost of the chip increases significantly as the number of the pins increases. This causes trouble, because the chip has to be able to communicate with its environment. For example, a picture database system has to be able to communicate with users through displays, secondary storages, and other computer systems. The path to the environment is extremely narrow if the entire system is implemented on one chip.

Chip interconnection is a major problem. As for the internal processor interconnections, in order to achieve fast intercommunication and low complexity, several kinds of interconnections between processors are proposed as surveyed by Haynes [HAY82]. As for the processor to processor intercommunication, several processor-memory interconnection network architectures are proposed. An example of the interconnection between processors and memories is as follows [AGR82].

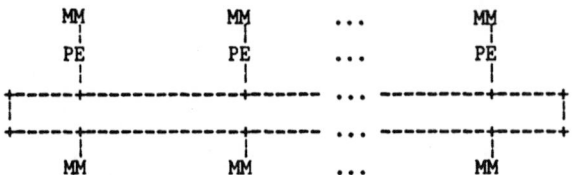

MM: Memory Unit
PE: Processing Element

The interconnection between multiple chips is more tight than that of the interconnection networks, while the interconnection is not so tight as the processor to processor interconnection on one chip. This level of interconnection architecture has not yet been investigated in full. We call this interconnection an inter-chip interconnection. In this paper, we assume that some kind of intercommunication between chips is present and the details of the communication are left to be worked out separately.

3.1. SPEC: Set Processing Execution Chip

In order to perform the extended relational calculus of PICCOLO, each chip is designed to perform all or part of the calculus. Each chip can perform the calculus internally if the data to be processed can be accommodated in the chip and the required data are already present in the chip. If the data to process cannot be hold in one chip, several chips are employed together. In the following, the function of the SPEC is specified by presenting the syntax to describe the program of the SPEC. In section 3.3, the extended relational calculus defined in section 2 is shown to be realizable on these SPECs. The internal architecture of the SPEC is illustrated in section 4.

3.2. Functions of the SPEC

The SPEC is designed to process the relational calculus by performing operations in parallel. The synchronization scheme of the SPEC is indebted to the mechanism based on the petri-net [PET], and the scheme is refined and given a programming language syntax named PCDL by Iizawa [IZW82]. The language syntax of PCDL is based on that of PASCAL [JEN75] with a few extensions in order to let the PCDL describe the concurrency and parallelism.

Extension 1. Concurrency control

Some data storages have flags to show whether the storage is empty or not. To control this capability, the assignment statement is extended as below.

$$\begin{array}{c} le \\ lf \end{array} \!\!\!\diagdown\!\!\!\diagup \begin{array}{c} := \end{array} \!\!\!\diagdown\!\!\!\diagup \begin{array}{c} re \\ rf \end{array}$$

The dot le means "If the left storage is empty, the operation is performed."
The dot lf means "If the operation is finished, the left storage is filled."
The dot rf means "If the right storage is full, the operation is performed."
The dot re means "If the operation is finished, the right storage is emptied."

Extension 2. Parallel processing

Statements enclosed by curly brackets are carried out in parallel.

This is a further extended version of the PCDL. Because the PICCOLO is based on the "set at a time" parallelism, the simple parallelism of the PCDL suffices for supporting the PICCOLO operation.

3.3. Translation from the Extended Relational Calculus to the PCDL of SPECs

The extended relational calculus proposed in section 2 can be supported by the SPECs. We show the transformation rule from the extended relational calculus to the PCDL of the SPECs. There are two extremes in this translation of the extended relational calculus to the PCDL. One is that SPECs are programmed before the operation begins. This case is called preprogramming. Another is that SPECs are programmed dynamically as the evaluation proceeds. This case is called deferred programming. In order to support the generic representation, deferred programming has to be allowed, while preprogramming is honored for the purpose of efficiency. The optimal combination of preprogramming and deferred programming depends on system configurations and the calculus to be evaluated. In the following the preprogramming is explained.

For explanation's purposes, the term $C:=\{(x,z)|(x,y)\in A^{\wedge}(y,z)\in B\}$ is used as a sample term. First, the chip allocation and data path have to be described. In order to represent the chip allocations and data paths, we describe them in a diagram called an evaluation diagram. The evaluation diagram is a directed graph whose node corresponds to chips which are not necessarily distinct. An arc corresponds to the inter-chip communication. We will show the method used to convert the term to the evaluation diagram.

(1) Construct a tree of the term.

The result of the sample term is the following diagram.

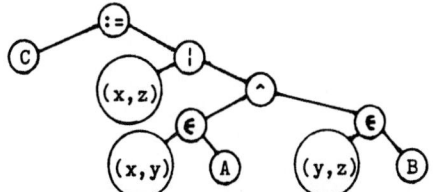

(2) According to the following rules the above tree is converted to a diagram.

In this diagram, the following notations are used.

A means SPECs which contain the set A.

G,H,... mean subdiagrams.

X,Y,... mean strings of variables.

x,y,... mean variables.

T and F mean the true and false values, respectively.

"eval" means that the program is activated when a variable is replaced with the input value.

A variable on an arc means that the variables are to be transferred on the arc in the direction of the arc.

1. Variable

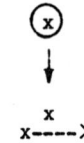

```
         x
      x----->
```

2. Function

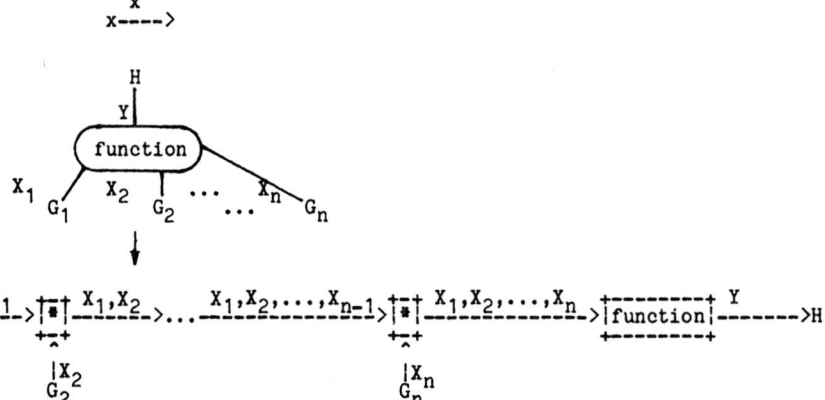

The function $|*|$ is a function to combine two or more input data.

2'. Function "="

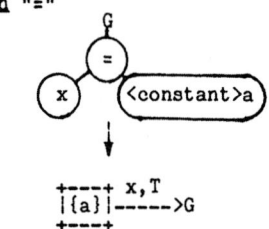

```
      +----+  x,T
      |{a}|----->G
      +----+
```

2". Function ":="
Case a)

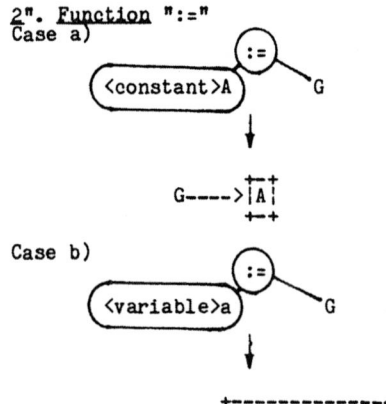

```
      G----->|A|
             +-+
```

Case b)

```
                 :=
      <variable>a      G


           a   +-----------------+
      ----->|eval(G--->|a|)|
                        +-+     |
            +-----------------+
```

2"'. Function "|"

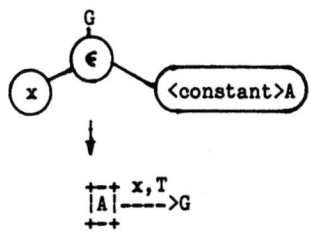

2"". Function "."

3. Function "∈"
Case a)

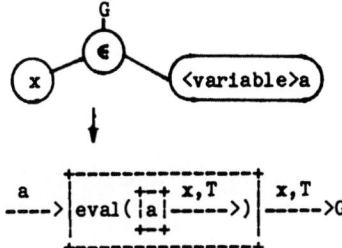

Case b)

4. Generic tuple
Case a)

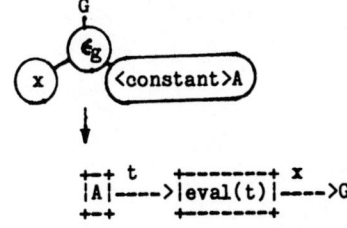

Case b)

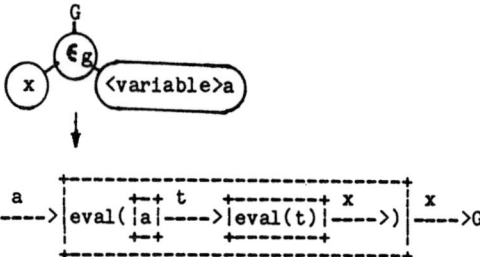

$$\begin{array}{c} a \\ ---> \end{array} \text{eval(|a|----> |eval(t)|--->)) |} \begin{array}{c} x \\ --->G \end{array}$$

5. Quantifier

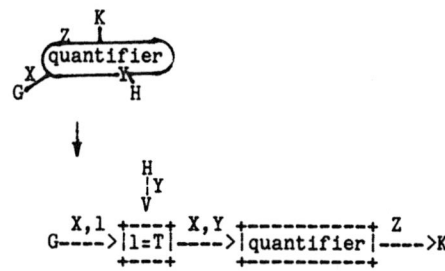

$$G \xrightarrow{X,1} \boxed{1=T} \xrightarrow{X,Y} \boxed{\text{quantifier}} \xrightarrow{Z} K$$

The result of the sample term is the following diagram.

$$\boxed{A} \xrightarrow{x,y,T} \boxed{*} \xrightarrow{x,y,T,y,z,T} \boxed{\char`\^} \longrightarrow \boxed{x,z} \xrightarrow{x,z} \boxed{C}$$

$$\bigg| y,z,T$$

$$\boxed{B}$$

(3) <u>Then the values to be transmitted are decided by the following steps.</u>

1. Redundant variable reduction

$$\begin{array}{ccc} X & & X' \\ ----> & \ldots & ----> \end{array}$$

If all the descendants use only a set of variables X', then $X = X'$. In this case, care must be take for the following diagram.

$$\boxed{A} \xrightarrow{x} $$

Suppose that $x=(x_1,x_2,\ldots,x_n)$. When the positions $\{i,j,\ldots,k\}$ of the tuple are not necessary from rule 1, then the tuple is represented by $(x_1,\ldots,x_{i-1},x_{i+1},\ldots,x_{k-1},x_{k+1},\ldots,x_n)$ to show which column is omitted. This notation is used when the positions of variables are of importance.

The result of the sample term is the following diagram.

$$\boxed{A} \xrightarrow{x,y,T} \boxed{*} \xrightarrow{x,T,z,T} \boxed{\char`\^} \longrightarrow \boxed{x,z} \xrightarrow{x,z} \boxed{C}$$

$$\bigg| y,z,T$$

$$\boxed{B}$$

2. By utilizing the properties of functions, several optimizations are possible.

A.

```
x,1,x',1' +-+ x",T
---------->|^|----->
          +-+
```

```
        ↓
```

```
x,T,x',T +-+ x",T
---------->|||----->
          +-+
```

B.

```
x',1 +--+ x
----->|x||--->
      +--+
```

```
    ↓
```

```
x,T +--+ x
----->|x||--->
      +--+
```

C.

```
x +--+ x
--->|x||--->
    +--+
```

```
    ↓
```

```
x
--->
```

D.

```
x,T,y,T +-+ x,y
---------->|^|---->
          +-+
```

```
     ↓
```

```
x,T,y,T +---+ x,y
---------->|xy||----->
          +---+
```

3. A constant value does not have to be transmitted through an arc. 3he programming of the chip which receives the constant is modified to use the constant value instead of the received value.

The result of the sample term is as follows.

```
+-+ x,y  +-+ x,z  +-+
|A|---->|■|---->|C|
+-+      +-+      +-+
          ^
         |y,z
         +-+
         |B|
         +-+
```

The programming of the SPEC is as follows.

1. Retrieve

```
+-+ x
|A|---->
+-+
```

Read out an element of the set A. In the PCDL, the program will look as follows, supposing that the data are transferred through a number n_of_ports of ports:

```
element: array[1..max_n_of_element] of record ... end;
```

```
n_of_ports: integer;
n_of_elements: integer;
i: integer;

program readout();
begin
    for i :=: 1 to n_of_elements div n_of_ports do
    { CONTROL_PORT ::: element[(i-1)*n_of_ports+j]; | j=1..n_of_ports }
    { CONTROL_PORT ::: element[(n_of_elements-1)-(n_of_elements-1) reminder
            'n_of_ports+j] | j=1..(n_of_elements-1) reminder n_of_ports}
end.
```

2. Function/quantifier

```
  X    +--------------------+  Y
----->|function/quantifier|---->
       +--------------------+
```

Evaluate the function or quantifier by using the input values. The program is analogous to the previous one.

3. Store

```
  X    +-+
----->|A|
       +-+
```

The set A is initialized and the elements which are input are stored in the relation. The program is analogous to the previous one.

The correspondence between the actual chip and the node of the evaluation diagram is not necessarily one-to-one. Any chip which contains an element of a relation A belongs to a node |A| of a diagram. It is true that more than one inter-chip connection may be allotted to one arc. Also it is true that more than one arc may be allotted to one inter-chip connection. If the initial node and the terminal node have the same chip in common, an interconnection is not actually established. Any chip can be assigned to the function or quantifier node if the chip can perform the operation.

3.4. Some Features of the SPEC

3.4.1. Multiple Copies of One Relation

In order to increase the efficiency of the PICCOLO database multiple copies of one relation may be present within the network to reduce the time consuming network data transfer. This mechanism is used also to reduce the data transfer time for programming chips. This is possible because the program is also a record of a relation in PICCOLO and thus the same mechanism can be used for data and programs.

3.4.2. Automatic Chip Allocation

The SPEC can be programmed to automatically allocate chips if the chip cannot hold all the elements of the relation. The essential part of this programming is as follows:

```
var
message: record
            m_destination: address;
            m_command: message_command;
            m_address: address;
```

```
          m_data: data;
      end;
  procedure alloc();
  var destination: address;
  begin
    message.m_destination = BROAD_CAST;
    message.m_command = FIND_FREE_CHIP;
    CONTROL_PORT := message;
    message =: CONTROL_PORT;
    if message.m_command = FOUND then
      begin
        message.m_destination = message.at_address;
        message.m_command = INITIALIZE;
        message.m_data = set_name;
        CONTROL_PORT := message;
        for ..... (* almost same as the read-out program in this section *)
        message.m_data = EOD;
        CONTROL_PORT := message;
      end
    else error(NO_MORE_SPACE);
  end;
```

By transferring half of the elements contained in the chip into the new chip, the B-tree by [BAY72] can be produced by employing a large number of SPECs.

3.4.3. Virtual Storage Handling

Data stored in a chip may be swapped out to other free chips or disks if necessary. Also the program of a chip may be swapped out if the program space of the chip is exhausted. The management of swap-in/swap-out is carried out by the SPEC program which will never swap out itself.

3.4.4. Error Resistance

If the number of chips in the system is very large, it becomes very difficult to assume the error rate of the chips to be negligible. To make the situation better, the system must be designed to be error resistant. The above SPEC architecture can be made error resistant by constructing a redundant evaluation diagram.

For example, the sample evaluation diagram can be modified to be evaluated more reliably as shown below.

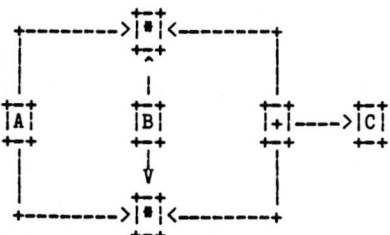

The function |+| checks the values from the top and the bottom to match. If the values do not match, an error is reported.

Because the top half evaluation diagram and the bottom half evaluation diagram work in parallel, the speed degradation is very small.

3.4.5. High availability

Because great freedom exists for assigning work to SPECs, it is easy to detach the malfunctioning chips without causing any serious harm to the system operation. The responsibility of the adjacent chip to the malfunctioning chip is to notify the chip which accesses the malfunctioning chip of the fact that the chip malfunctions; this forces the requesting chip to find alternate chips.

3.4.6. Concurrency processing

3.4.6.1. Sharing a chip by multiple diagrams

If more than one diagram has to share the same chip, then the chip generates a unique number for the identification of the evaluation diagram. By the identification number which accompanies the input data, the related evaluation diagram is identified and the right program is activated for processing the data in the shared chip.

3.4.6.2. Locking data

The locking of the data is already done because each data processing operation is carried out by a distinct evaluation diagram which can be identified by each chip. A more elaborate locking scheme can be supported by combining primitive locking schemes.

```
+-+      +-+      +-+
|A|---->|C|<----|B|
+-+      +-+      +-+
 a                b
```

This type of evaluation diagram might be erroneous, because the values generated by the evaluation diagram a or b are never used. This kind of mistake can be detected by programming the SPEC to monitor the reference, and modification of its data.

3.4.7. Authorization and protection

The view of a database system can be supported by the modification of the operation to the SPECs. For example, suppose that a read out request to relation A from a user "a" means that "a" reads a set A' which is a generic relation. If A' contains a generic tuple (x :name|(x :name) A), then the relation A for the user a is the restriction of the relation A to the attribute "name". This management is carried out by the chip which contains the relation A. When the relation A chip is requested to perform a read-out operation by user "a", the chip searches the possible user list for the relation A. If the user has no authorization to use the relation A, then an error is reported to the requester. If the user has defined the view by generic tuples as explained above, the request is forwarded to the right relation A' according to the view.

3.5. Evaluation of the SPEC

We briefly evaluated the speed and the logic cost of the SPEC architecture for the sample operation. The possible optimizations of the SPEC architecture are also shown with the modification of the diagram. The time required to evaluate the given calculus depends greatly on whether or not the whole operation is carried out in one SPEC, or not. An evaluation within one SPEC is called an _internal evaluation_. An evaluation in which each node of the diagram corresponds to different SPECs is called an _external evaluation_. The time required to process an operation by multiplying the processing SPECs is called a _multiple evaluation_. Each estimation is marked by (internal), (external), or (multiple) to indicate on which assumption the evaluation is carried out. Here, the estimation of the time and the logic is very rough. The purpose of this estimation is to tune the functional design in this level of designs. A more elaborated estimation is obtained when the internal architecture of the SPEC is determined. In the following, several parameters are introduced.

Parameters:

t_X: time required to transfer a data through an arc identified by X.

t_f: time required to perform the function f.

p_f: the number of processors to process the function f in one SPEC.

n_A: the number of elements in a relation A.

$f[a_1\text{->}b_1, a_2\text{->}b_2, \ldots, a_n\text{->}b_n]$: the term obtained by replacing all the occurrences of a_1 to b_1, a_2 to $b_2, \ldots, a_n$ to b_n in the term f.

l_f: the amount of logic required to perform the function f.

$n_A|(,y)$: the number of elements (,y) in a relation A for a fixed y.

t_{setup}: chip programming time

Assumption:

A network communication initiation and termination overheads are negligible.

The number of elements of a relation is very large.

3.5.1. Relational Join Example

A. Extended relational calculus

$C := \{(x,z) \mid (x,y) \in A, (y,z) \in B\}$

B. Evaluation diagram

```
+-+  xy   +-+  xz   +-+
|A|---->|*|---->|C|
+-+  X   +-+  Z   +-+
              ^
           Y|yz
            +-+
            |B|
            +-+
```

C. PCDL program

The programming of |*| is to load the set A first, and then compare them with all the elements of set B.

D. Time estimation

$T_e \sim n_A{}^*t_X + n_C{}^*t_Z + n_B{}^*(t_Y + (n_A/p_=^{}{}^*t_=))$ (external)

$T_i \sim n_A * n_B / p_= * t_=$ (internal)

$T_m \sim Te[n_A \to n_A / p_X, n_B \to n_B / p_Y]$ (multiple)

Modification A: The elements of sets A and B are sorted according to field y.

C. PCDL program

The programming of $|*|$ is modified to utilize the fact that field y is sorted.

D. Time estimation

$T_e \sim n_A * t_X + n_B * t_Y + (n_A + n_B) * t_< + n_C * t_Z + \max(t_{sort(A)}, t_{sort(B)})$ (external)

$T_i \sim (n_A + n_B) * t_< + \max(t_{sort(A)}, t_{sort(B)})$ (internal)

E. Additional logic

l_{sort} (per chip)

Modification B: The SPEC has an associative memory capability.

B. Evaluation diagram

```
+-+ x,y +-+ x,z +-+
|A|---->|B|---->|C|
+-+ X   +-+ Y   +-+
```

C. PCDL program

The programming of chip B is changed to find the element (y,z) by giving the value of y.

D. Time estimation

$T_e \sim n_A * (t_X + n_B / p_{assoc} * t_{assoc} + \Sigma_{(,y) \in A} n_B | (y,) * t_Z)$ (external)

$T_i \sim n_A * n_B / p_{assoc} * t_{assoc}$ (internal)

E. Additional logic

$l_{assoc} * p_{assoc}$ (per chip)

3.5.2. Deferred Programming Example

A. Extended relational calculus

$B := \{(x) | (a) \in A, (x) \in a\}$

B. Evaluation diagram

```
+-+ a    +----------------+   x   +-+
|A|----->|eval( |a|----->)|----->|B|
+-+ X    |      +-+ Y     | Z     +-+
         +----------------+
```

C. PCDL program

The "eval" is the built-in function of the SPEC used for programming the chips according to this diagram.

D. Time required

$T_e \sim n_A * (t_X + t_{setup}) + \Sigma_{a \in A} n_a * (t_Y + t_Z)$ (external)

$T_i \sim n_A * t_{setup} + \Sigma_{a \in A} n_a * t_Y$ (internal)

$$T_m \sim n_A/p_a * (t_{setup} + max_{i=1} \quad \Sigma_{n \in A_i} n_a * (t_Y + t_Z)) \text{ (multiple)}$$

3.5.3. Function unit distribution

Some function units such as addition, multiplication, floating point arithmetic, and sorting processing require so much logic that it is too costly to provide each chip with these units. Thus, there is a trade-off between the processing speed and the logic required. In the functional specification, the following two cases are identified.

Case 1. Each SPEC contains more than one function unit.

Case 2. Some SPEC contains the function unit, while other SPECs do not contain it.

The effect of the second case is that the internal evaluation might not be possible. The performance of the system is estimated by the number P_u of function units in the chip.

3.5.4. Optimization by physical data structure

In the specific chips, the physical data structure is utilized to speed up the execution of the data operations. In the PICCOLO scheme, the physical data structure is not explicitly specified by the command to the system. The external functional specification of the SPEC imposes no restriction on the physical data structure. Thus, it is possible to employ a specific data structure to speed up execution. An example is a sorted set as used in example 3.5.1. Another example is to use a storage address in a chip as a tuple id. The management of this operation is carried out by the SPEC which contains the relation. Thus, from outside the tuple id looks like it actually exists.

3.5.5. Summary

In this section, we will show some typical picture processing operations and the roughly estimated time required to process the operations. Also, the improvements of the processing time by giving the SPEC more logic are listed.

1. The increment of gray levels in a two-dimensional picture.

Assume that a two-dimensional picture is represented as a tuple (id,v) in a relation A.

1.A. Extended relational calculus

$A' := \{(id,v') | (id,v) \in A, v' = v+1\}$

1.B. Evaluation diagram

```
    +--+ id,v  +--+ id,v' +--+
    |A |------>|+1|------->|A'|
    +--+  X    +--+  Y     +--+
```

1.D. Time estimation

$T_e \sim n_A * (t_X + t_{+1} + t_Y)$ (external)

$T_i \sim n_A * t_{+1}/p_{+1}$ (internal)

$$T_m \sim n_A/p_{+1}{}^*(t_X+t_{+1}+t_Y) \text{ (multiple)}$$

2. Relationship Join.

A join of two relations through a relationship is named a **relationship join**.

2.A. Extended relational calculus

$$B:=\{(v_a,v_b)\,|\,(id,a,id_a,b,id_b)\in A,(id_a,v_a)\in a,(id_b,v_b)\in b\}$$

2.B. Evaluation diagram

2.D. Time estimation

$$T_e \sim n_A{}^*t_X+n_A{}^*t_{setup}+n_A{}^*t_Y+n_A{}^*t_{setup}+n_A{}^*t_Z+\Sigma_{a\in n_A} n_a{}^*t_=/p_=+\Sigma_{b\in n_B} n_b{}^*t_=/p_= \text{ (external)}$$

$$T_i \sim n_A{}^*\Sigma_{a\in n_A} n_a{}^*t_=/p_=+n_A{}^*\Sigma_{b\in n_A} n_b{}^*t_=/p_=+n_A{}^*t_{setup}+n_A{}^*t_{setup} \text{ (internal)}$$

$$T_m \sim \max_i(n_A/p_A{}^*t_X+n_A{}^*t_{setup}+n_A/p_A{}^*t_Y+n_A/p_A{}^*t_Z+n_A/p_A{}^*t_{setup}$$

$$+\Sigma_{a\in A_i} n_a{}^*t_=/p_=+\Sigma_{b\in A_i} n_b{}^*t_=/p_=) \text{ (multiple)}$$

2. Modification A: Associative memory

2.D. Time estimation

$$T_e \sim n_A{}^*t_X+n_A{}^*t_{setup}+n_A{}^*t_{assoc}+n_A{}^*t_Y+n_A{}^*t_{setup}+n_A{}^*t_Z+n_A{}^*t_{assoc} \text{ (external)}$$

3. Rotation of a two-dimensional pixel array

Assume that a picture is represented as a tuple (x,y,v) in a relation A and the picture is rotated by a rotation matrix M.

3.A. Extended relational calculus

$$A':=\{(x,y,v')\,|\,(x',y')=(x,y)^*M,v'=h_1{}^*v_1+\dots+h_4{}^*v_4,$$

$$x_1=|x'|,y_1=|y'|,h_1=(x'-x_1)^*(y'-y_1),(x_1,y_1,v_1)\in A,\dots\}$$

Here, $|x|$ is an integer closest to a real number x.

3.B. Evaluation diagram

In this diagram, subdiagrams for calculating v_3 and v_4 are omitted for the sake of simplicity.

This "matrix*" function is implemented by the following diagram:

```
   x,y  x  +----+
   ----+--->|*a11|---+
        |   +----+   |
        |            V
        |  y +----+  +-+  x'  +-+  x',y'
        +--->|*a21|->|+|----->|*|-------->
        |   +----+   +-+      +-+
        |                     ^-+
        |  x +----+  +-+  y'   |
        +--->|*a12|->|+|-------+
        |   +----+   +-+
        |            ^-+
        |  y +----+   |
        +--->|*a22|---+
            +----+
```

Here, the function "*" outputs a set of data by combining the input data.

3.D. Time estimation

$T_e \sim n_A{}^*(\max(t_X, t_{matrix^*}, t_Y) + n_A{}^* t_Z + n_A{}^* t_=/p_= + t_{mult} + t_{add})$ (external)

$T_i \sim n_A{}^*(t_{matrix^*} + n_A{}^* t_Z + n_A{}^* t_=/p_= + t_{mult} + t_{add})$ (internal)

$T_m \sim T_e[n_A \to n_A/p_X]$ (multiple)

$t_{matrix} \sim t_* + t_+$

4. Eight neighborhood operations on a two-dimensional pixel array.

We assume that a two-dimensional pixel array is represented as a tuple (x,y,v) in a set A.

4.A. Extended relational calculus

$A':=\{(x,y,v') | v'=f(v1,v2,\ldots,v8),(x-1,y-1,v1)\in A,\ldots,(x+1,y+1,v8)\in A\}$

4.B. Evaluation diagram

```
+-+ x,y,v   x,y  +------+ x-1,y-1  +-+   x,y,v1  +-+
|A|--------+---->|-1,-1|--------->|=|<--------1-|A|
+-+ X      | Y   +------+ Z       +-+   W        +-+
           .                       | x,y,v1
           .                       |
           .                       | x,y,v1,v2,v3,v4,v5,v6,v7
                                   V
                                  +-+
                                  |+|
                                  +-+
                                   | x,y,v1,v2,v3,v4,v5,v6,v7,v8
                                   V
                                  +-+ x,y,v   +-+
                                  |f|-------->|A'|
                                  +-+ T       +-+
```

4.D. Time estimation

$T_e \sim n_A{}^*(t_X + t_Y + t_- + t_Z + t_= + n_A{}^*(t_W + t_=/p_=) + t_V + t_S + t_X + t_f + t_T)$ (external)

$T_i \sim n_A{}^*(t_- + t_= + n_A{}^* t_=/p_= + t_f)$ (internal)

$T_m \sim T_e[n_A \to n_A/p_X]$ (multiple)

Modification A: Associative memory

$T_e \sim n_A{}^*(t_X + t_Y + t_- + t_Z + t_{assoc} + t_V + t_S + t_f + t_T)$ (external)

5. Filling a region in a two-dimensional pixel array

We assume that a two-dimensional pixel array is represented by a tuple (x,y,v) in a relation A; the boundary pixels of the region have the value one and the rest of the pixels have the value zero.

5.A. Extended relational calculus

$B:=\{(x,y,v'')|(x,y,v)\in A,(x,y+1,v')\in A,v''=v*(1-v')\}$

$C:=\{(x,y,v'')|v''=max(sum(\{v|(x,y',v)\in B,y'<y\})\ mod\ 2,v'),(x,y,v')\in B\}$

5.B. Evaluation diagram

```
+-+ x,y,v'                            x,y,v'                +-----+ x,y,v"  +-+
|B|--------+----------------------------------->|=,max|-------->|C|
+-+ X      |x,y              Y                   +-----+ Z       +-+
           |                                        ^
|x',y,v    |T                                       |
           V                                        |
           +-+ x',y,v +------+ y,v   +------+ y,v    |
+--------->|<|-------->|adder|------>|mod 2|-------->+
+--------+ +-+ U       +------+ V    +------+ S
  W                                   
```

5.D. Time estimation

$T_e \sim n_B*(n_B*(t_</p_<+t_{adder}+t_{mod2}+t_=/p_=)+t_{max}$
$\quad +max(t_X+t_Y+t_Z,t_T+t_U+t_V+t_S,t_W+t_U+t_V+t_S))$ (external)

$T_i \sim n_B*(n_B*(t_</p_<+t_{adder}+t_{mod2}+t_=/p_=)+t_{max})$ (internal)

$T_m \sim T_e[n_B->n_B/p_X]$ (multiple)

Modification A: Associative memory

$T_e \sim n_B*(n_B*(t_</p_<+t_{adder}+t_{mod2})+t_{max}$
$\quad +max(t_X+t_Y+t_Z,t_T+t_U+t_V+t_S,t_W+t_U+t_V+t_S))$ (external)

4. Internal control structure of the SPEC

Each SPEC contains a large number of processing elements (PE for short) organized in some geometry. Each PE contains a processor, local data storages, and programs to be executed. The internal control structure of the PEs is organized as a tree as shown below.

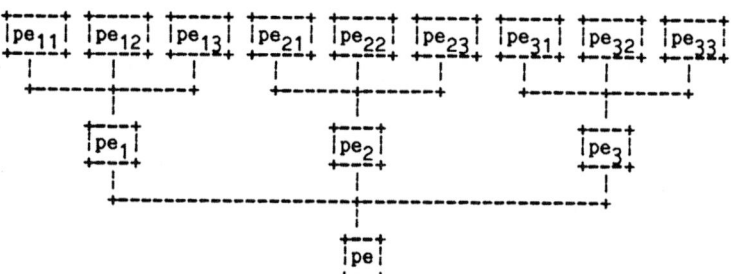

The PE is identified as $pe_{i_1,i_2,i_3,...,i_1}$ as above, and the degree of the tree in level m is denoted as degree(m). $pe_{i_1,i_2,...,i_1}$ controls its child PEs $pe_{i_1,i_2,...,i_1,i_{1+1}}$ $(1<=i_{1+1}<=degree(1))$. Several tree structure multiprocessor machines such as the X-Tree [PAT79] take a binary tree structure as not only a control structure but also as a data path. In our scheme, the geometry of the data path is not yet limited to a tree structure in this level of decisions. This control structure is suitable for supporting the PCDL as shown below. By increasing the number of interconnection paths in one chip, the performance of the chip increases, while the logic which is required for one chip also increases. So this is a trade-off between the performance and the logic required. This is also true for the communication interfaces with the outside. Some of the PEs are allowed to communicate with the outside. The communication interface through which a PE can communicate with the outside is called a port denoted as PO.

4.1. Programming the PE

The programming of the SPEC explained in the previous section is decomposed into multiple programs of PEs in the SPEC. Each PE's program controls itself and its child PE's program execution.

A. Concurrency control

An empty flag is attached to a storage to show the emptiness/fullness of the storage.

B. Parallel processing

$$\{p_1;p_2;\ldots;p_n\}$$

If this operation is assigned to a PE $pe_{i_1,\ldots,i_l}$, then $pe_{i_1,\ldots,i_l,1},\ldots,$ $pe_{i_1,\ldots,i_l,n}$ are programmed to do the operations $p_1, p_2,\ldots,p_n$, respectively. Assume that these programs are activated under the name of "p". Then, the $pe_{i_1,\ldots,i_l}$ will do the following operations to process the statement.

```
call("p",1,2,...,n);
wait(1,2,...,n);
```

Here, the "call" operation starts the operation named "p" at its children 1 through n. Then, it waits until the end of these operations.

4.2. Example

The function $|*|$ used in section 3.5 is carried out by PEs as follows.

Program of pe:

```
procedure cmp();
begin
  call("load",1,...,n);
  wait(1,...,n);
  while true do
    begin
      message =: portY;
      if message.m_data = EOD then break;
      data = message.m_data;
      call("cmp",1,...,n);
      wait(1,...,n)
    end
end;
```

Program of pe[i]:

```
procedure load();
begin
  leng = 0;
```

```
      while true do
        begin
          message =: portX[i];
          if message.m_data = EOD then break;
          pe[i,leng] = message.m_data;
          leng = leng+1
        end
  end;

  procedure cmp();
  begin
    for j = 1 to leng do
      begin
        if pe[i,j].y = data.y then
          portZ := pe[i,j].x | data.z
      end
  end;
```

The speed is slightly increased by giving each input/output port queues to buffering data. This kind of improvement is hidden from users at this level of abstraction.

5. Conclusion

In this paper, the VLSI technology application for a picture database computer is explained. The final step of the architecture engineering to interface the logic design is again divided into several steps. In these steps, the modularity of the design is very important to gradually tailor the logical design to the chip design. The estimations as shown in this paper allow us to make architectural decisions in the early stage of design. The intercommunication protocol and internal data communication of the SPEC are not investigated in this paper, but are left for future work.

References

[AHM82] H. M. Ahmed, J. M. Demosme, and M. Morf, "Highly Concurrent Computing Structures for Matrix Arithmetic and Signal Processing," IEEE Computer, Jan., 1982, pp. 66-82.

[ALE78] N. Alexandridis, and A. Klinger, "Picture Decomposition Tree Data-structures, and Identifying Directional Symmetries as Node Combinations," Computer Graphics and Image Processing, 8, 1978, pp. 43-77.

[AND80] M. Andrew, "Principle of Firmware Engineering in Microprogramming Control," Computer Science Press, 1980.

[BAY72] R. Bayer, and E. M. McCreight, "Organization and Maintenance of Large Ordered Indexes," Acta Informatica, 1, 1972, pp. 173-189.

[CAR] E. Carlson and P. Manty, Tech. Reports from IBM San Jose Research Laboratory.

[CHA77] S. K. Chang, J. Reuss, and B. H. McCormick, "An Integrated Relational Database System for Pictures," Proc. IEEE Workshop on Picture Data Description and Management (PDDM), IEEE Computer Society, 1977, pp. 49-60.

[COD70] E. F. Codd, "A Relational Model of Data for Large Shared Data Banks," Comm. ACM 13, 1970, pp. 377-387.

[COD72] E. F. Codd, "Relational Completeness of Data Base Sublanguages," in Data Base Systems, ed. R. Rustin, 1972, Prentice-Hall, pp. 65-98.

[FU79] N. S. Chang and K. S. Fu, "Query-by-Pictorial-Example," Proc. COMPSAC 79, Chicago, November, 1979, pp. 325-330.

[GUZ81] A. Guzman, "A Heterarchical Multi-Microprocessor Lisp Machine," Proc. of IEEE Workshop on Computer Architecture for Pattern Analysis and Image Data Management (CAPAIDM), Hot Spring, Virginia, 1981, pp. 309-317.

[HAN75] P. B. Hansen, "Concurrent PASCAL," IEEE Trans. Software Eng., SE-1, 1975, pp. 199-207.

[HAY82] L. S. Haynes, R. L. Siewiorek, D. P. Siewiorek, and D. W. Mizell, "A Survey of Highly Parallel Computing," IEEE computer, Jan., 1982, pp. 9-24.

[HOP79] J. E. Hopcroft, and J. D. Ullman, "Introduction to Automata Theory, Language, and Computations," Addison-Wesley, 1979.

[JEN75] K. Jensen, and N. Wirth, "PASCAL User Manual and Report," second edition, Spring-Verlag, 1975.

[KAN82] K. Kanasaki, K. Yamaguchi, and T. L. Kunii, "A Software Development System Supported by a Database of Structures and Operations", COMPSAC 82, Nov., 1982, pp. 343-350.

[KEN81] J. T. Kenneth, and A. F. Havey, "Tutorial: Local Computer Networks," second edition, IEEE computer society press, 1981.

[KRU56] J. B. Kruscal, "On the shortest Spanning Subtree of a Graph and the Travelling Salesman Problem," Proc. Amer. Math. Soc., 7, 1956, pp. 48-50.

[KUN74] T. L. Kunii, T. Amano, H. Arisawa, and S. Okada, "An Interactive Fashion Design System INFADS," Proc. of Conference on Computer Graphics and Interactive Techniques, 1974.

[KUN75] T. L. Kunii, S. Weyl, and J. M. Tenenbaum, "A Relational Data Base Schema for Describing Complex Picture with Color and Texture," Proc. of Second International Joint Conference on Pattern Recognition, IEEE Computer Society, Aug., 1975, pp. 310-316; reprinted in Policy Anal. Info. 1, No.2, 1978, pp. 127-142.

[OHB79] N. Ohbo, K. Shimizu, and T. L. Kunii, "A Graph-theoretical approach to region detection," Proc. IEEE Computer Society's Third International Computer Software & Applications Conference, Chicago, Nov., 1979, pp. 751-756.

[PAT79] D. A. Patterson, E. S. Fehr, and A. H. Sequin, "Design Consideration for the VLSI processor of X-Tree," Proc. Ann. Symp. Computer Architecture, 1979, pp. 90-100.

[RAM77] C. V. Ramamoorthy, and H. H. So, "Software Requirements and Specifications: Status and Perspective," Appendix A to Requirements Engineering Research Recommendations, Aug., 1977.

[RAM82] C. V. Ramamoorthy, and Y. W. Ma, "Impact of VLSI on Computer Architecture," VLSI Electronics 3 ed. by Norman G. Einspruch, Academic press, 1982, pp. 1-23.

[RIE80] C. Rieger, et al., "ZMOB: A Highly Parallel Multiprocessor," Proc. of Picture Data Description and Management (PDDM), Aug., 1980, pp. 298-304.

[ROE79] R. P. Roesser, "Two-Dimensional Microprocessor Pipelines for Image Processing," IEEE Transactions on Computers, Vol. C-27, No. 2, Feb., 1978, pp. 144-156.

[SIE79] H. J. Siegel et al., "An SIMD/MIMD Multiprocessor System for Image Processing and Pattern Recognition," Proc. of 1979 Conference on Pattern Recognition and Image Processing, Aug., 1979, pp. 214-220.

[YAM80] K. Yamaguchi, N. Ohbo, T. L. Kunii, H. Kitagawa, and M. Harada, "ELF: Extended relational model for Large, Flexible picture databases," Proc. IEEE Workshop on Picture Data Description and Management (PDDM), Asilmar, California, Aug., 1980, pp. 95-100.

[YAM81] K. Yamaguchi and T. L. Kunii, "Logical Framework of a Picture Database Computer," Proc. of IEEE Workshop on Computer Architecture for Pattern Analysis and Image Data Management (CAPAIDM), Hot Spring, Virginia, Nov., 1981, pp. 284-292.

[ZLO75] M. M. Zloof, "Query by Example," Proc. AFIPS, NCC, Vol. 44, 1975, pp. 431-438.

ADVANCES IN DIGITAL IMAGE PROCESSING FOR DOCUMENT REPRODUCTION

P. Stucki
IBM Zurich Research Laboratory
8803 Rueschlikon
Switzerland

ABSTRACT

The properties of conventional, ordered dot-pattern generation techniques for bi-level halftone representation are examined and compared with the properties of error-diffusion-based, disordered dot-pattern-generation algorithms. The various processing steps necessary for adaptation of the disordered halftone pattern-generation technique to digital image hardcopy reproduction with non-ideal computer-output printing devices are described. It includes procedures for spatial distribution of thresholding errors, suppression of dot-density artifacts and compensation for dot overlap. These procedures represent the core of a Multiple-Error Correction Computation Algorithm called MECCA, the objective of which is to linearize the non-ideal printing process in order to minimize the loss or shift of tonal gradations. Finally, the performance of MECCA is compared with a conventional digital screening technique, and the various reproduction-quality versus computational-complexity trade-offs are discussed.

INTRODUCTION

The current trend in computer industry is towards smaller, faster and cheaper computers. The most important impetus for this increase in computer power has come from the improved performance of Large-Scale Integration (LSI) electronic circuitry which has brought switching speed into the pico-second range. Very Large-Scale Integration (VLSI) technology with its capability to further reduce the cost of storage and computing will strongly impact the speed and scope of the evolution from today's data-processing equipment to the information-processing systems of the future.

Today, the field of conventional data processing and data transmission is well established. While text processing and electronic document distribution have already been introduced, image and voice acquisition, processing, storage and retrieval are still in their infancy. However, here too, new technology is on the way to help. Advanced generations of computers and progress in the area of I/O devices will allow remote copying of facsimiles, storage and retrieval of financial, legal and technical documents, handling of speech files, etc.

As more and more text-processing systems are able to cope with the arrangement of text within a page, attention shifts to the problem of digital preparation of artwork and illustrative material so that it can be composed along with text. The convenience of giving editors not only the option of placing a picture in its correct position, but also cropping, sizing and altering the photographic quality of these pictures, maximizes the amount of editorial control over the final product. Thus, the intent of such a publishing system is not only to set artwork and illustrative material, but also to process, store and retrieve them upon demand.

Artwork and illustrative material generally consists of logotypes, line-art and photographs. Their reproduction with an all-point-addressable computer-output printer requires them to be in bi-level form since printing processes generally exhibit only two levels of optical density: The presence or absence of ink or toner particles, for example. While the transformation of logotypes and line-art into bi-level form consists of a straightforward thresholding operation, the conversion of an analog continuous-tone image into a sampled bi-level halftone representation has motivated a number of research efforts.

The purpose of this paper is to describe an improved method to generate and manipulate bi-level halftone representations for high-resolution achromatic image rendition.

CONVERSION OF A CONTINUOUS-TONE IMAGE INTO A BI-LEVEL HALFTONE REPRESENTATION

General

For the conversion of a continuous-tone image into a bi-level halftone representation, a photograph or a natural scene is first scanned using a Charge-Coupled Device (CCD) or a video camera, for example. The resulting analog signal is then translated into digital form by sampling and amplitude quantizing the different shades of gray, ranging from all white to all black. The resolution of the digitized continuous-tone image is a function of the number of lines scanned N, the number of discrete samples or picture elements (pel) taken M, and the number of signal amplitude levels quantized Q.

A straightforward approach to convert a continuous-tone image into a bi-level halftone representation is to simulate the process of halftone photography. The latter is used in lithography and photoengraving and, in principle, works as follows: When a halftone negative is made from a continuous-tone copy, a halftone screen - in the case of a contact screen an out-of-focus pattern of different dots and corresponding spaces - is placed in the light path between the camera lens and a sheet of high-contrast photographic film. The basic function of the halftone screen and the high-contrast photographic film is to convert the intermediate tones of the original continuous-tone copy into solid dots of equal density but varying size, the centers of which are placed on a regular grid.

Digital Halftone Photography

In digital halftone photography, the function of screening is achieved by comparing at each picture element position x, the scanned and digitized continuous-tone signal $q(x)$ against a variable level, over i pels repetitive, threshold or dither pattern $t[mod(x,i)]$, to make a black or white decision $h(x)$ [1,2]. Figure 1 shows the principle of variable-level thresholding in one dimension. Corresponding examples of bi-level representations are shown in Figure 2.

In order to break-up the width-modulated vertical line structure originating from variable-level thresholding in one dimension, the continuous-tone signal $q(x)$ has to be compared at each picture-element position x against a variable threshold profile that varies in the two image dimensions x and y (Figure 3).

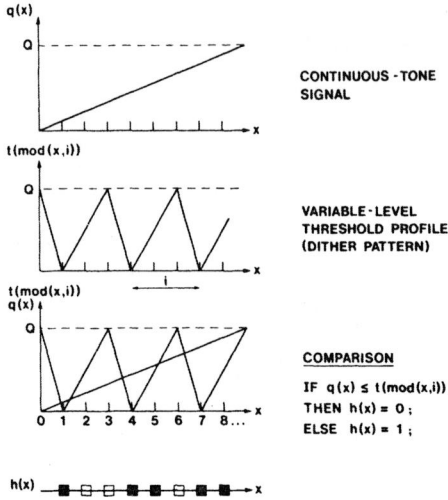

Figure 1. Principle of variable-level thresholding.

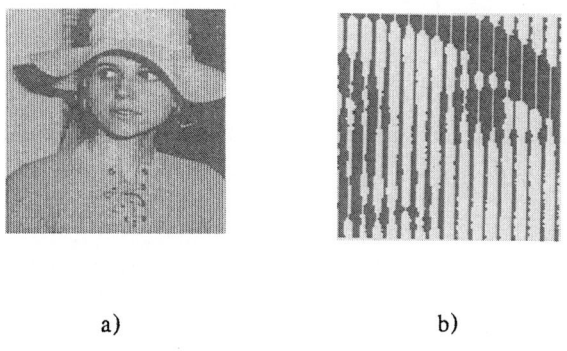

a)   b)

Figure 2. One-dimensional variable-level thresholding.

    a) Portrait picture
       N=600 lines, M=600 pel.
    b) Cropped and magnified eye portion of portrait picture
       N=600 lines, M=600 pel.

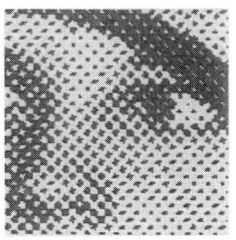

a)   b)

Figure 3. Two-dimensional variable-level thresholding.

     a) Portrait picture
       N=600 lines, M=600 pel.
     b) Cropped and magnified eye portion of portrait picture
       N=600 lines, M=600 pel.

In practice, the digitized continuous-tone signal $q(x,y)$ is compared against a two-dimensional, variable-level, over i and j repetitive threshold or dither pattern $t[\text{mod}(x,i), \text{mod}(y,j)]$, to generate a bi-level halftone representation $h(x,y)$, such that for $q(x,y) \leq t[\text{mod}(x,i), \text{mod}(y,j)]$, $h(x,y) = '0'$ (white) and '1' (black) otherwise. Because of the repetitive nature of the threshold profile $t[\text{mod}(x,i), \text{mod}(y,j)]$ used, this screening method leads to a so-called *ordered dot pattern bi-level halftone representation*.

Threshold-Profile Optimization

The penalty paid for the improved halftone rendition by two-dimensional variable-level thresholding is a certain loss of spatial resolution or pictorial detail. Over the past years, several design approaches to optimize the threshold profile with respect to spatial resolution and tonal fidelity have been proposed [3,4]. Typically, an optimized threshold profile for best-possible spatial-resolution rendition consists of a set of threshold values as shown in Figure 4a).

22	6	18	2	21	5	17	1
14	30	10	26	13	29	9	25
20	4	24	8	19	3	23	7
12	28	16	32	11	27	75	31
21	5	17	1	22	6	18	2
13	29	9	25	14	30	10	26
19	3	23	7	20	4	24	8
11	27	15	31	12	28	16	32

19	25	23	17	14	8	10	16
21	31	29	27	12	2	4	6
28	30	32	22	5	3	1	11
18	24	26	20	15	9	7	13
14	8	10	16	19	25	23	17
12	2	4	6	21	31	29	27
5	3	1	11	28	30	32	22
15	9	7	13	18	24	26	20

a)   b)

Figure 4. Optimized threshold profiles $t(i,j)$ for $Q=32$ amplitude quantization levels with respect to:

a) spatial resolution; b) tonal fidelity.

This arrangement of threshold values can be characterized as one in which the $t(i,j)$th and the $t(i,j)+1$th threshold values are spatially located as far as possible away from one another in the threshold profile. Digital screens fulfilling this characteristic generate *dispersed dot patterns* having the capability of reproducing high spatial resolution at a linear dot-area-coverage versus amplitude-level relationship when reproduced with an ideal, i.e., square dot, on/off printing device. However, when using a non-ideal, i.e., circular dot, on/off printing device producing overlapping dots, as in non-impact ink-jet or electrophotographic printing, the dot-area-coverage versus amplitude relationship becomes dot-pattern dependent, i.e., non-linear and therefore, difficult to control without losing tonal fidelity.

A natural way to reduce the darkening effect of dot overlap is to generate *clustered dot patterns*. Typically, an optimized threshold profile for best-possible tonal-fidelity rendition consists of a set of threshold values as shown in Figure 4b). In this case, the threshold profile can be characterized as one in which the $t(i,j)$th and the $t(i,j)+1$th threshold values are spatially arranged as closely as possible to one another, and where the clustered dot patterns run under 45° and 135°, respectively, an angle at which the cut-off bandwidth for binocular vision is reduced by approximately 10 % as compared to horizontally and vertically oriented gratings [5]. Dot overlap is naturally compensated at the interior of the dot clusters and a close-to-linear dot-area-coverage versus amplitude-level relationship is achieved at the expense of a certain loss in spatial resolution. An analytic procedure to design clustered dot patterns based on the concept of SUPER-CIRCLES, or dot clusters with envelope of the general form

$$x^{2k} + y^{2k} = r^{2k},$$

$$\text{where} \quad 1 \leq k \leq \infty ,$$

is described in [6].

The Rendition of Scanned Logotypes, Line-Art and Photographs

Artwork and illustrative material to be processed consists of logotypes, line-art and photographs or a combination of these. It is essential that the reproduction device be capable of rendering these different type of images with adequate quality. Logotypes and line-art can be characterized by sharp boundaries and the scanned and digitized continuous-tone signal $q(x,y)$ contains fast transients. Its proper rendition calls for high spatial-resolution requirements while a moderate number of amplitude quantization levels is sufficient to guarantee adequate reproduction quality. Photographs are quite different from logotypes and line-art. In areas with slowly-varying gray levels, the issue of tonal fidelity is very important and fine amplitude quantization is required, otherwise contouring will appear. At the same time, moderate spatial-resolution requirements usually guarantee adequate rendition quality.

A method to improve the digital bi-level representation whenever logotypes, art-work and photographs are intermixed consists of computing the Laplace gradient Δ at every picture element position (x,y) of the digitized continuous-tone signal $q(x,y)$ and adaptively switching between two different, two-dimensional threshold profiles. For example, if a fast transient is detected, $q(x,y)$ will be compared with a threshold profile generating dispersed dot patterns, otherwise the threshold profile rendering clustered dot patterns will be used [7]. The concept of adaptive switching of dispersed and clustered dot patterns for bi-level halftone representation removes, to a certain extent, the spatial-resolution and tonal-fidelity restrictions encountered in conventional screening.

Multiple-Copying of Bi-Level Halftone Representations

The various bi-level halftone representation techniques discussed so far have been evaluated in terms of their multiple digital copying potential and as a function of different scanning-aperture and light-collector device characteristics. The investigation is based on an extended simulation to model the various steps involved in the entire halftone document copying cycle, starting off from a digitized continuous-tone signal $q(x,y)$ ready for conversion into its first bi-level halftone representation $h(x,y)$, followed by a re-scanning process performed under arbitrary tilt angle α to obtain a new digitized continuous-tone signal $q'(x,y)$ ready for a second, not necessarily the same conversion into a new bi-level halftone representation $h'(x,y)$, etc.

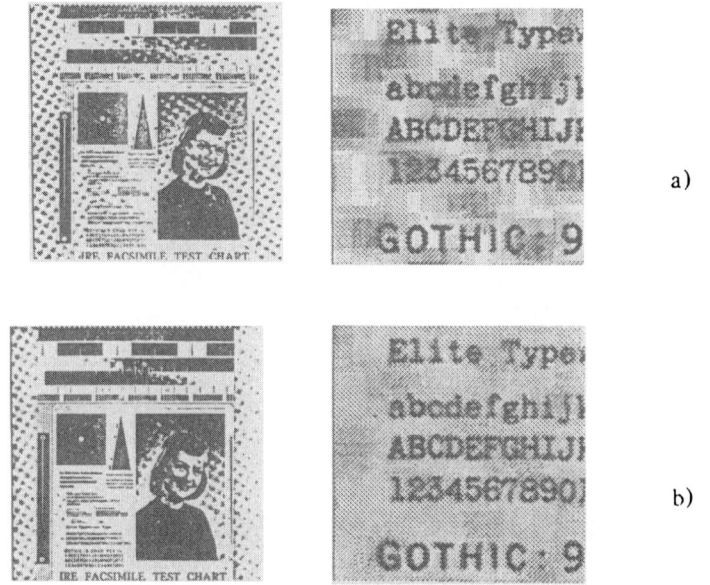

a)

b)

Figure 5. Re-scanning bi-level halftone representations.

Clustered dot pattern re-scanned with: a) a rectangular-shaped aperture profile; b) a Gaussian-shaped aperture profile.

Figure 5a) shows a clustered dot pattern bi-level halftone representation originally printed at a resolution of 400 pel/inch and re-scanned at a resolution of 480 pel/inch, a tilt angle of $\alpha = 1.5^{\circ}$ and assuming a square aperture profile, which is found typical for CCD's. The same picture re-scanned under the same condition, but assuming a Gaussian aperture profile as encountered in laser scanners is shown in Figure 5b). The Moiré pattern formation observed in Figures 5a) and b) are a direct function of the generating dot-pattern structure, the tilt angle α and the shape of the aperture profile. A possible way to suppress the formation of such interferences is to enlarge the scanning aperture beyond the screening resolution of the original picture. However, such a filtering approach degrades the sharpness of the re-scanned picture. Figures 6a), and b) show the result of Moiré-free re-scanning using an enlarged Gaussian aperture profile.

a)

b)

Figure 6. Moiré-free re-scanning using an enlarged Gaussian aperture profile.

Re-scanning with: a) 8 pel x 8 pel aperture;
b) 12 pel x 12 pel aperture.

Pseudo-Random Thresholding

An alternative way to suppress the formation of Moiré patterns is to use pseudo-random thresholding techniques to generate bi-level representations of halftones. In this case, the digitized continuous-tone signal $q(x,y)$ is compared, at each pel position (x,y), with a pseudo-random threshold signal $t(x,y)$ to generate $h(x,y)$ such that for $q(x,y) \leq t(x,y)$, $h(x,y) = '0'$ (white) and '1' (black) otherwise. Because of the pseudo-random nature of the threshold signal $t(x,y)$, this screening method leads to a so-called *disordered dot-pattern bi-level halftone representation* (Figure 7).

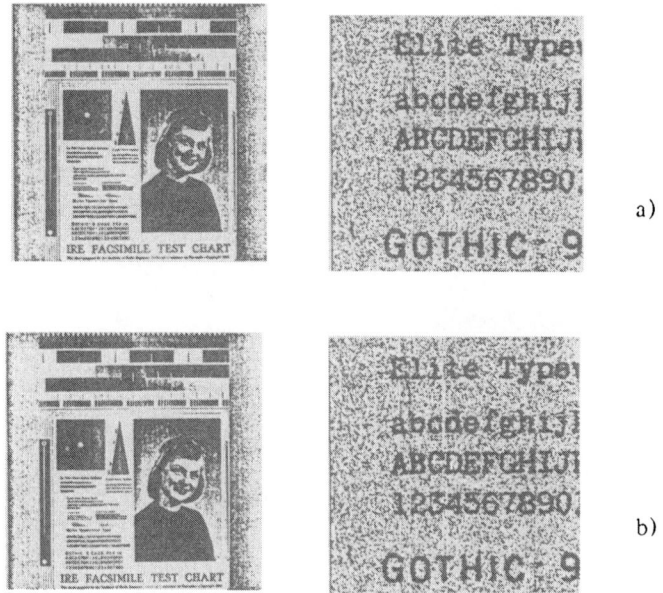

a)

b)

Figure 7. Pseudo-random thresholding to generate bi-level representations of halftones.

a) Original; b) re-scanned with 8 pel x 8 pel Gaussian aperture.

It is interesting to note that no Moiré pattern formation takes place when re-scanning and subsequent pseudo-random thresholding a bi-level representation originally generated with pseudo-random thresholding halftone techniques [Figure 7b)]. Unfortunately, the actual appearance of the picture quality obtained is rather poor. An explanation for the shortcoming of this approach is the fact that the pseudo-random threshold signal $t(x,y)$ has a uniform spectral density distribution, i.e., it contains equally strong components at all spatial frequencies which add-up to the spectral density distribution proper to the pictorial information in the visible as well as in the invisible frequency range. Pseudo-random thresholding can be classified as a *stochastic, non-linear process* capable of transforming a digitized continuous-tone signal $q(x,y)$ into a disordered pattern bi-level halftone representation.

Fixed-Level Thresholding and Error Diffusion

In order to avoid the undesired uniform spectral density distribution characteristics of pseudo-random thresholding, a *deterministic, non-linear process* leading to a disordered dot-pattern bi-level halftone representation is based on constant-level thresholding and error-carry (Figure 8).

Its principle of operation in one dimension is as follows: At every picture-element location x, a corrected digital continuous-tone value $q'(x)$, representing the sum of the digitized continuous-tone signal $q(x)$ and a carried error value $e(x-1)$ arising from processing the previous neighbor pel, is compared with a spatially-independent constant-level threshold $t(x)$ set at $Q/2$, to generate a bi-level halftone representation $h(x)$ such that for $q'(x) = q(x) + e(x-1) \leq Q/2$, $h(x) = $ '0' (white) and 'Q' (black) otherwise.

A straightforward one-dimensional error-carry approach tends to generate subjectively disturbing line texture and, in practice, the error value $e(x)$ is distributed amongst the surrounding neighbors. This two-dimensional process is called error diffusion. It requires more processing power and buffer memory than the error-carry approach but has the advantage of removing to a large extent the spatial-resolution and tonal-fidelity restrictions encountered in ordered-pattern halftone generation. There exist basically two different strategies to achieve error-diffusion in two dimensions. In a first approach, fractions of the error value $e(x,y)$ at each picture-element position (x,y), are apportioned to the neighborhood digitized continuous-tone signal values still to be thresholded [8]. In a second approach, a weighted average of previously determined neighborhood error values is added to the continuous-tone signal value $q(x,y)$ at picture-element position (x,y), ready for thresholding [9]. This second approach is called minimized average error algorithm, and was developed to display continuous-tone pictures on gas-discharge displays.

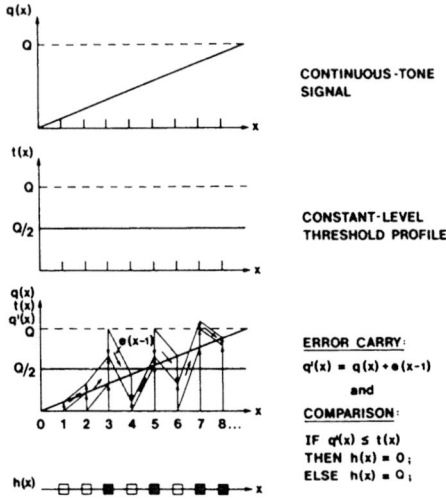

Figure 8. Principle of constant-level thresholding and error carry.

The purpose of this paper is to describe how to enhance the latter concept of error diffusion to generate pseudo-disordered dot patterns for bi-level halftone representation and how to adapt this technique for *achromatic hardcopy reproduction with non-ideal printing devices.*

PSEUDO-DISORDERED DOT-PATTERN GENERATION FOR
BI-LEVEL HALFTONE HARDCOPY REPRODUCTION

Constant-Level Thresholding and Two-Dimensional Error Diffusion

The principle of constant-level thresholding and two-dimensional error
diffusion is shown in Figure 9.

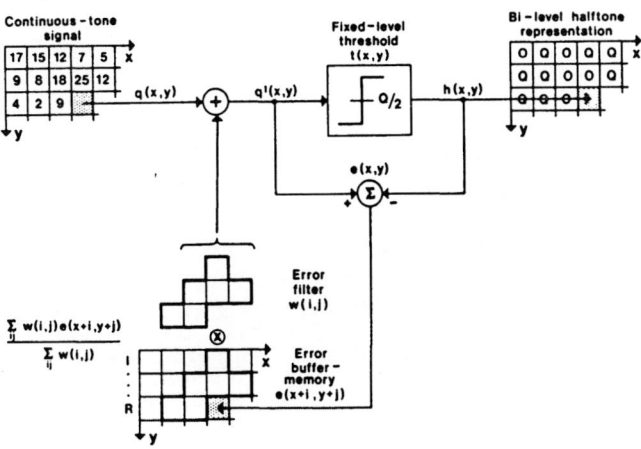

Figure 9. Principle of constant-level thresholding and two-
dimensional error diffusion.

At each picture-element position (x,y), a corrected digital continuous-
tone signal q'(x,y), representing the sum of the digital continuous-tone signal
q(x,y) and a weighted average of previously computed errors, is first calculated:

$$q'(x,y) = q(x,y) + \sum_{i,j} (w(i,j)\, e(x{+}i, y{+}j)) \;/\; \sum_{i,j} w(i,j),$$

where $w(i,j)$:   error-filter impulse response
                          extending over (i,j),

   $e(x{+}i, y{+}j)$:   error value at $(x{+}i, y{+}j)$.

The corrected digital continuous-tone signal $q'(x,y)$ is then compared with a fixed-level threshold $t(x,y)$ set at $Q/2$ to generate a halftone representation $h(x,y)$ such that for $q'(x,y) \leq Q/2$, $h(x,y) = $ '0' (white) and 'Q' (black) otherwise. Although the outcome of this non-linear operation is either '0' or 'Q', the actual density of the digitized continuous-tone signal $q(x,y)$ can be anywhere in between. To complete the computation cycle, the thresholding error $e(x,y) = q'(x,y) - h(x,y)$ is determined and loaded into an S x M wide error buffer memory for subsequent use in the following thresholding cycles. Constant-level thresholding and two-dimensional error diffusion for the computation of intermediate average densities proceeds from left to right and from top to bottom in the N x M array.

Filter Optimization

Constant-level thresholding and two-dimensional error-diffusion produces a deterministic arrangement of disordered dot patterns. The 'degree of disorder' is determined by the characteristics of the error filter used. In order to determine the optimum filter size and filter coefficients, three different error-filters with impulse responses shown in Figure 10 were studied.

```
                                        1    2 2 2 2 2 2
1            2                 1   1 2 4 2 1      .    2 4 4 4 4 4 2
           2 8 2               .   2 4 8 4 2      .    2 4 8 8 8. 4 2
S    2 8 x                     S   4 8 x          S    2 4 8 x

     a)                            b)                       c)
```

Figure 10. Impulse response coefficients $w(i,j)$ of different-size
           error filters.

           a) 'Small' spatial extent (S=3).
           b) 'Medium' spatial extent (S=3).
           c) 'Large' spatial extent (S=4).

It should be noted that the values of all weight coefficients are chosen to be 2^k, where k = 0, 1, 2, 3, ... , in order to improve computational efficiency. The 'small' error filter is capable of reproducing dark areas reasonably well. For the rendition of light areas, however, the 'degree of disorder' in the dot-placement structure is not satisfactory. Instead, their positions are correlated in such a way as to produce diagonal streaks in the bi-level halftone representation. The 'degree of disorder' can also be expressed in terms of a two-dimensional power-spectrum plot which clearly indicates a certain angular preference in the distribution of the individual dot-pattern-frequency components [Figure 11a)].

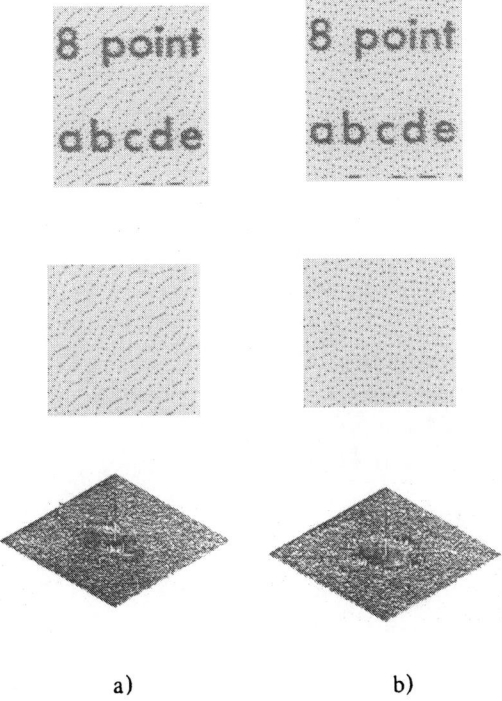

a)   b)

Figure 11. Error-filter size and 'degree of disorder'.

a) 'Small' spatial extent; b) 'medium' spatial extent.

It should be noted that these streaks correspond to the direction of maximum weighting in the error filter, i.e., the two eights run diagonally next to each other. Disturbing texture can be avoided - at the expense of computational efficiency - by increasing the spatial extent of the error filter. The rendition of light areas is very much improved when using the 'medium' error filter. In this case, the two-dimensional power-spectrum plot no longer indicates an angular preference, but shows a close to rotation-symmetric distribution of the individual dot-pattern-frequency components [Figure 11b)]. Experiments using the 'large' error filter indicate no substantial improvement in terms of dot-placement regularity.

Look-Ahead Fitering to Reduce Dot-Density Artifacts

Because of the finite spatial extent and the band-limiting characteristics of the impulse response w(i,j) of the error filter, subjectively disturbing 'density overshoots' occur when reproducing high-contrast, large-area patterns (Figure 12).

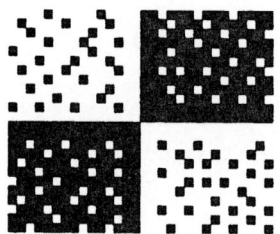

Figure 12. Example of 'density overshoot' in chequer-board pattern.

In order to suppress the formation of such artifacts, the constant-level thresholding and two-dimensional error-diffusion process has been complemented with a two-dimensional transient detection or look-ahead filter operating over a T x M wide input value buffer memory (Figure 13).

At each picture element position (x,y), the corrected digital continuoustone signal q'(x,y) is now determined as a sum of the digital continuous-tone signal q(x,y), a weighted average of previously computed errors and, new, a term reflecting the local image texture:

$$q'(x,y) = q(x,y) + \sum_{i,j} (w(i,j)\, e(x+i,y+j)) \,/\, \sum_{i,j} w(i,j)$$

$$+ \sum_{i,j} (z(k,l)\, q(x+k,y+l)) \,/\, \sum_{i,j} z(k,l),$$

where $z(k,l)$: look-ahead filter impulse response extending over (k,l),

$q(x+k,y+l)$: amplitude value at $(x+k,y+l)$.

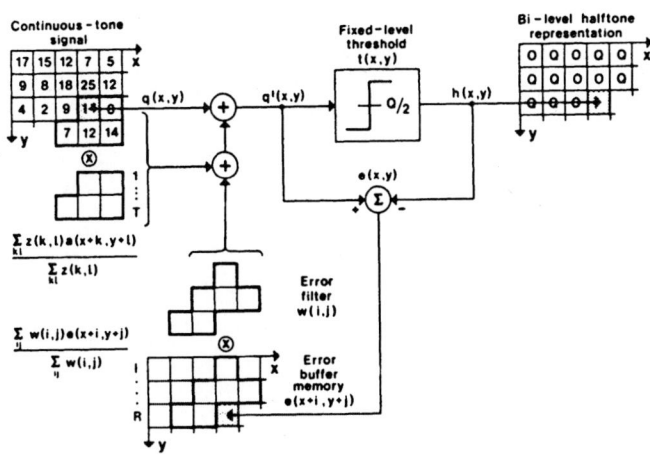

Figure 13. Principle of constant-level thresholding, two-dimensional error diffusion and look-ahead filtering.

Like in the previous case, the corrected digital continuous-tone signal $q'(x,y)$ is then compared with a fixed-level threshold $t(x,y)$ set at Q/2 to generate a halftone representation $h(x,y)$ such that for $q'(x,y) \leqslant Q/2$, h(x,y) = '0' (white) and 'Q' (black) otherwise. To complete the computation cycle, the error e(x,y) = q'(x,y) - h(x,y) is again determined and loaded into the S x M wide error buffer memory for subsequent use in the following threshold cycles.

Several look-ahead filters have been investigated whereby the impulse response $z(k,l)$ shown in Figure 14a) represents a compromise. Together with a 'medium' error filter, this look-ahead filter prevents the appearance of artifactual density overshoots when reproducing high-contrast, large-area patterns without noticeable blurring [Figure 14b)].

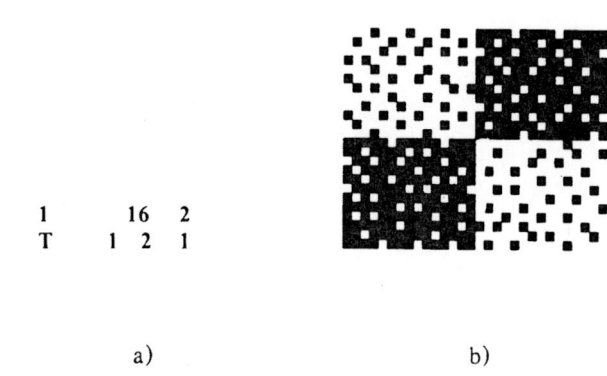

```
1      16   2
T      1  2  1
```

a)   b)

Figure 14. Look-ahead filtering to prevent density overshoots.

    a) Impulse response z(k,l).
    b) Reproduced chequer-board pattern.

Dot-Size Correction to Compensate for Dot Overlap

The degree of perfection reached in constant-level thresholding, two-dimensional error diffusion and look-ahead filtering is not yet sufficient to perform well for hardcopy reproduction with non-ideal printing devices.

In all-point-addressable computer-output printing, electrophotographic, ink-jet and other non-impact printing technologies are used. Square-shaped dots, as shown in the simulated chequer-board patterns of Figures 13 and 14a), cannot be printed with toner and ink-based recording technologies. Instead, individual dots are circular in shape when printed on ideal paper.

Given this technology limitation, the reproduction of continuous areas of black is only possible by causing adjacent dots to overlap [Figure 15a)]. In this case, however, the exact rendition of average densities becomes dependent on the combination of adjacent neighboring dots already printed, and the resulting overdarkening has to be removed with a space-variant dot-size correction algorithm.

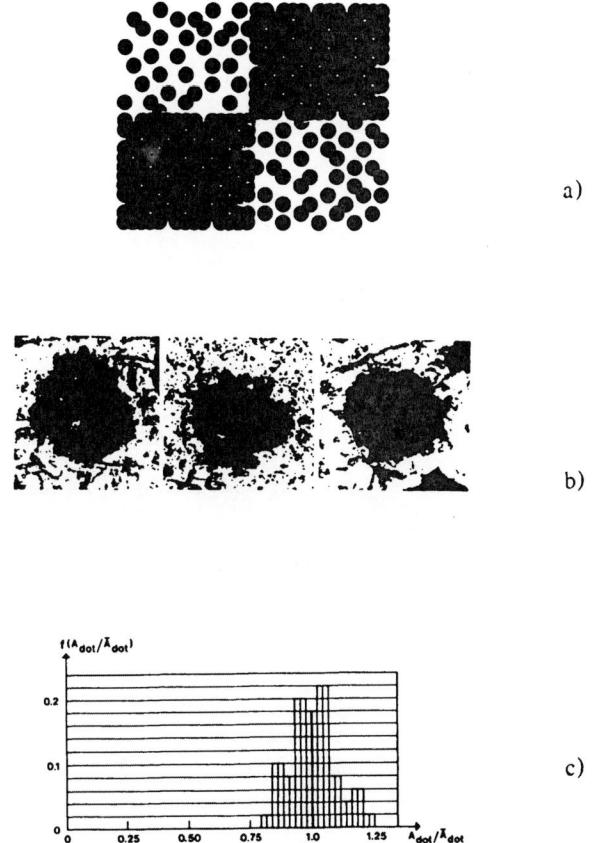

a)

b)

c)

Figure 15. Printing with circular-shaped dots.

    a) Reproduced chequer-board (simulation).
    b) Examples of dots printed on electro-sensitive paper.
    c) Histogram of measured dot size A_{dot}.

In practice, paper does not exist in ideal form, and fluctuations in dot shape as shown in Figure 15b) are very likely to occur. Figure 15c) shows the histogram of measured dot size A_{dot}. It was determined by taking microscopic photographs of a set of dots and measuring their surface with a Polarplanimeter. Dot-placing experiments on normal and electro-sensitive paper have shown that the fluctuations in dot size and shape are stochastic and that they do not influence the overall average densities over larger dot-pattern areas. Thus, the analytic treatment of dot-size correction then consists in determining the area of white paper which a circular dot newly covers, when it is printed (Figure 16).

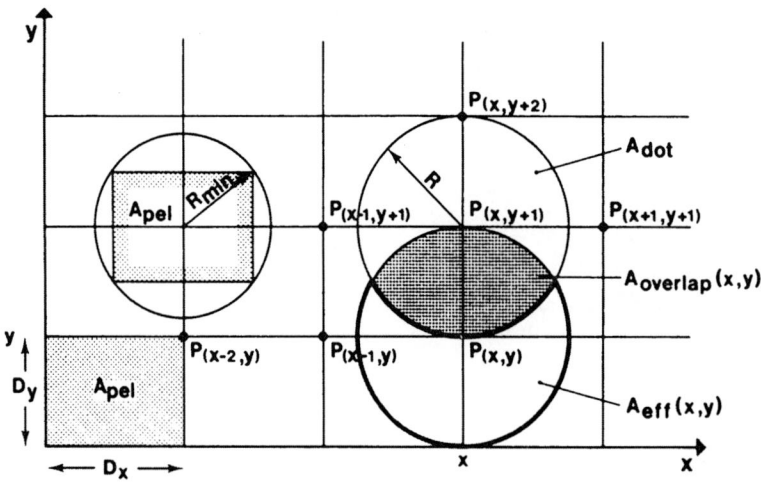

Figure 16. Parameters for dot-size correction.

$P(x,y)$ denotes the dot center for the current print position. Previously printed dots may be located at $P(x-1,y)$, $P(x-2,y)$, ..., $P(x+1,y+1)$, $P(x,y+1)$, $P(x-1,y+1)$, ..., $P(x,y+2)$, ..., etc.. D_x and D_y represent the distance between adjacent print positions in x and y. The area of a picture element is defined as

$$A_{pel} = D_x D_y.$$

In order to guarantee the complete coverage of A_{pel} with a circular dot, its radius R must fulfill the condition

$$R \geq R_{min} = (\sqrt{D_x^2 + D_y^2})/2.$$

Area overlap amongst adjacent circular dots with radius R occurs for all

$$R > |min(D_x, D_y)|/2.$$

Since

$$\sqrt{D_x^2 + D_y^2} > min(D_x, D_y),$$

area overlap will always occur under the above condition of full pel coverage whenever adjacent dots are present.

If $A_{dot} = \pi R^2$ denotes the area of a dot without overlap and $A_{overlap}(x,y)$ the total area of overlap amongst adjacent dots, the effective area $A_{eff}(x,y)$ of white paper newly covered by a dot printed at P(x,y) amounts to

$$A_{eff}(x,y) = A_{dot} - A_{overlap}(x,y)$$

$$= \pi R^2 - A_{overlap}(x,y).$$

Finally, the dot-size correction value $\lambda(x,y)$ applying at print position P(x,y), is defined as

$$\lambda(x,y) = A_{eff}(x,y)/A_{pel}.$$

If L represents the number of adjacent print positions considered, $\lambda(x,y)$ can take any of 2^L values depending on the combinations of printed dots. Given D_x, D_y, R and L proper to an all-point-addressable computer-output printer, all 2^L possible dot-size-correction values $\lambda(x,y)$ have to be determined and stored in a device-specific look-up table.

A straightforward approach to compute $\lambda(x,y)$ is to use raster-graphic techniques (Figure 17).

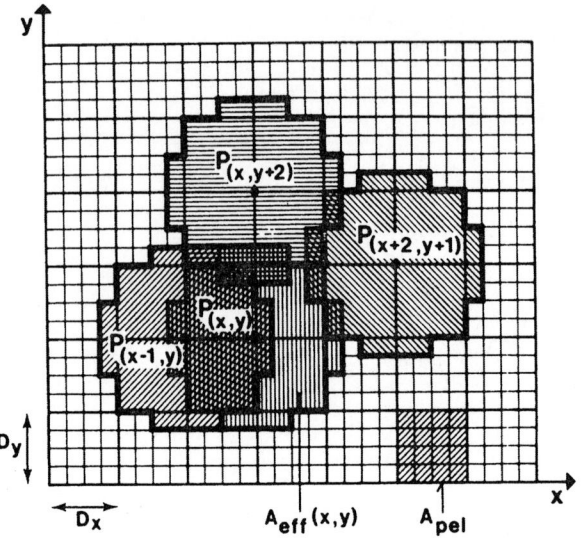

Figure 17. Raster-graphic approach to determine $A_{eff}(x,y)$.

For each of the 2^L possible dot-pattern combinations, digital approximations of printed circular dots are placed at relative print positions spaced D_x and D_y raster elements apart. Another digital approximation of a printed circular dot is then placed at $P(x,y)$, and the number of newly covered 'white' raster elements, representing $A_{eff}(x,y)$, determined. For a worst-case $\lambda(x,y)$ accuracy of 1%, A_{pel} should be represented by 10 x 10 raster elements or more.

The dot-size-correction factor $\lambda(x,y)$ can also be determined analytically. In order to fulfill the condition of covering A_{pel} with a circular dot, as well as to guarantee a unique dot-overlap order, computation of $A_{overlap}(x,y)$ and consequently $\lambda(x,y)$ places certain restrictions on the geometry and the values of D_x, D_y, R and L as outlined in the following two cases.

FIRST CASE (L=4)

In this case, dots printed at the four adjacent print positions $P(x-1,y)$, $P(x,y+1)$, $P(x-1,y+1)$ and $P(x+1,y+1)$ may contribute to area overlap when a new dot is printed at $P(x,y)$.

Restrictions on D_x and D_y

The condition to guarantee a fixed dot-overlap order implies that the largest distance between $P(x,y)$ and any of its L=4 neighbors

$$D_{max} = \sqrt{D_x^2 + D_y^2},$$

be less than the next-larger distance between $P(x,y)$ and any other print position possible

$$D_{next} = 2 \min(D_x, D_y).$$

Given this print-position geometry constraint, the restriction on the values of D_x and D_y can be expressed by

$$\max(D_x, D_y) \leq \sqrt{3} \min(D_x, D_y).$$

Restriction on R

In order to guarantee complete coverage of A_{pel} with a circular dot, as well as to exclude any overlay from all print positions other than the four adjacent neighbors, the range of R is determined by

$$(\sqrt{D_x^2 + D_y^2})/2 \leq R \leq \min(D_x, D_y).$$

Discussion

The selection of D_x, D_y and R within these boundaries allows the analytical computation of $\lambda(x,y)$. For $D_x = D_y = D$, the dot radius R can vary from 0.707D to D, while for $D_y = 3 D_x$, it cannot assume any value other than D_x.

The determination of all possible 2^4 $A_{overlap}(x,y)$, $A_{eff}(x,y)$ and $\lambda(x,y)$ values when printing a new dot at $P(x,y)$ is described in Appendix I.

SECOND CASE (L=6)

In this case, the four adjacent print positions $P(x-1,y)$, $P(x,y+1)$ $P(x-1,y+1)$ and $P(x+1,y+1)$ as well as the two next-closest print positions $P(x-2,y)$ and $P(x,y+2)$ may contribute to area overlap when a new dot is printed in $P(x,y)$.

Restrictions on D_x and D_y

The condition to guarantee a fixed dot-overlap order implies that the largest distance between $P(x,y)$ and any of its $L=6$ neighbors

$$D_{max} = 2 \max(D_x, D_y),$$

be less than the next-larger distance between $P(x,y)$ and any other print position possible,

$$D_{next} = \min(\sqrt{D_x^2 + 4D_y^2}, \sqrt{D_y^2 + 4D_x^2}).$$

Given this print-position geometry constraint, the restriction on the values of D_x and D_y can be expressed by

$$\max(D_x, D_y) = 2/\sqrt{3} \min(D_x, D_y).$$

Restriction on R

In order to guarantee complete coverage of A_{pel} with a circular dot, as well as to exclude any overlay from all print positions other than the six neighbors considered, the range of R is determined by

$$\max(D_x, D_y) \le R \le \min[(\sqrt{D_x^2 + 4D_y^2})/2, (\sqrt{D_y^2 + 4D_x^2})/2].$$

Discussion

The selection of D_x, D_y and R within these boundaries allows the analytical computation of $\lambda(x,y)$. For $D_x = D_y = D$, the dot radius R can vary from D to $1.11D$, while for $D_y = 2/3\, D_x$, it cannot assume any value other than D_y.

The determination of all possible 2^6 $A_{overlap}(x,y)$, $A_{eff}(x,y)$ and $\lambda(x,y)$ values when printing a new dot at $P(x,y)$ is described in Appendix II.

Dot-Size Corrected Bi-Level Representation of Halftone

In order to guarantee an overlap or dot-size corrected bi-level representation of halftones, the corrected digital continuous-tone signal $q'(x,y)$, determined as a sum of the digital continuous-tone signal $q(x,y)$, a weighted average of previously computed errors and a term reflecting the local image texture, is first computed. The value of $q'(x,y)$ is then compared with a fixed-level threshold $t(x,y)$ set at $Q/2$ to generate a halftone representation $h(x,y)$ such that for $q'(x,y) \leq Q/2$, $h(x,y) = '0'$ (white) and $'Q'$ (black) otherwise. To complete the computation cycle, a new overlay and dot-size dependent error $e(x,y) = q'(x,y) - [\lambda(x,y) h(x,y)]$ is then determined and loaded into the $S \times M$ wide error buffer memory for subsequent use in the following thresholding cycles.

The results obtained using a 'medium'-size error filter $w(i,j)$, the optimized look-ahead filter $z(k,l)$ as well as dot-size correction parameters $D_x = D_y = D$ and $R = 0.8D$ are shown in Figure 18.

 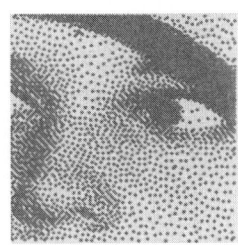

a)   b)

Figure 18. Constant-level thresholding, two-dimensional error diffusion, look-ahead filtering and dot-size correction.

   a) Portrait picture
      N=600 lines, M=600 pel.
   b) Cropped and magnified eye portion of portrait picture
      N=600 lines, M=600 pel.

For comparison, the same picture reproduced without dot-size correction is shown in Figure 19.

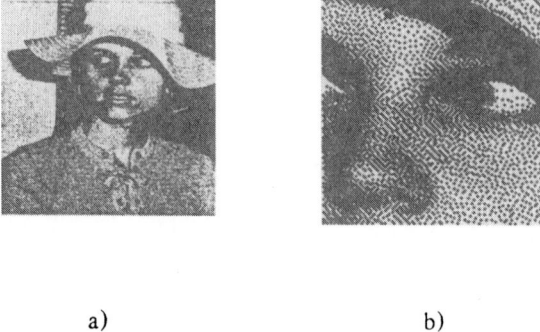

a)   b)

Figure 19. Constant-level thresholding, two-dimensional error diffusion and look-ahead filtering.

    a) Portrait picture
       N=600 lines, M=600 pel.
    b) Cropped and magnified eye portion of portrait picture
       N=600 lines, M=600 pel.

The area coverage c as a function of a synthetically generated amplitude quantization level q, varying over a range $0 < q < 127$, where '0' represents black and '127' white, is plotted in Figure 20. Curve a) corresponds to an 'open-loop' mode of the dot-size-correction scheme, i.e., the area coverage c is proportional to the newly covered area $A_{eff}(x,y)$ integrated over the entire range of g and without the dot-size-correction factor $\lambda(x,y)$ influencing the computation of the error value $e(x,y)$. Curve b) corresponds to a 'closed-loop' mode of the dot-size-correction scheme, i.e., the area coverage c is proportional to $A_{eff}(x,y)$ integrated over the entire range of q and with $\lambda(x,y)$ being used for the computation of $e(x,y)$. Working with exact values for D_x, D_y and R, the use of dot-size correction yields a linear correspondence between the area coverage c and the synthetically generated gray value q.

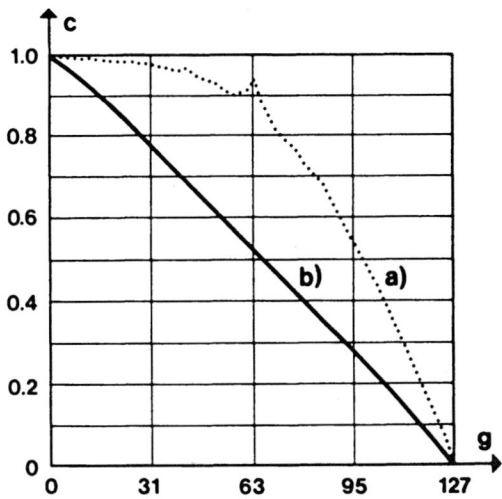

Figure 20. Area coverage c as a function of a synthetically generated amplitude quantization level q ('0' ≡ black, '127' ≡ white).

a) Without, b) with dot-size correction.

Several steps towards adapting the pseudo-disordered dot-pattern generation technique for bi-level halftone hardcopy reproduction have been discussed. They include procedures for spatial distribution of thresholding errors, the suppression of artifactual density overshoots and the compensation for dot overlap when using non-ideal computer-output printing devices. These procedures are all integrated into a single *Multiple-Error-Correction Computation Algorithm called MECCA.*

DISCUSSION OF MECCA

Comparison with SUPER-CIRCLE Screening

The Multiple-Error Correction Computation Algorithm combines linear and non-linear signal-processing techniques to yield a bi-level halftone representation that is, on a minimized average basis, directly proportional to the continuous-tone input value. Linearizing a non-ideal printing process with MECCA minimizes the loss or shift of tonal gradations and therefore, guarantees best-possible halftone rendition for given printer characteristics.

To verify *qualitatively* the performance of MECCA and SUPER-CIRCLE screening techniques, typical examples of continuous-tone text, graphics and natural images were processed. The illustrations used in these experiments are cropped pictures from the IEEE Facsimile Test Chart, originally scanned at a spatial resolution of 200 pel/inch and quantized into $Q=255$ amplitude levels. Prior to the actual MECCA and SUPER-CIRCLE processing, the cropped pictures were scaled to 750 x 750 pels using two-dimensional third-order polynomial interpolation and resampling techniques [6].

Continuous-tone text (Figure 21)

Due to the high spatial-resolution rendition property of MECCA, continuous-tone text, where edge information is known to be visually important, is reproduced with well-defined contours and without noticeable screening artifacts. In contrast, the visibility of impaired contours due to the clustering of dots in SUPER-CIRCLE screening, makes the latter a bad choice for computer-output printing of continuous-tone text. A cropped and magnified letter 'A' out of Figure 21 is reproduced in Figure 22. It clearly shows the poor reproduction of edges obtained with SUPER-CIRCLE screening.

Continuous-tone graphics (Figure 23)

Due to the pseudo-disordered dot-pattern-arrangement property of MECCA, continuous-tone graphics, where topological connectivity is known to be visually important, is reproduced with good definition and without noticeable screening artifacts. In contrast, the visibility of interferences or Moire patterns formed by the superposition of an ordered

Figure 21. Bi-level represenation of continuous-tone text.

    a) MECCA, Q=256.
    b) SUPER-CIRCLE, Q=32.
       Screening resolution: 70 clusters of dots/inch.
    c) SUPER-CIRCLE, Q=64.
       Screening resolution: 47 clusters of dots/inch.
    d) SUPER-CIRCLE, Q=128.
       Screening resolution: 35 clusters of dots/inch.

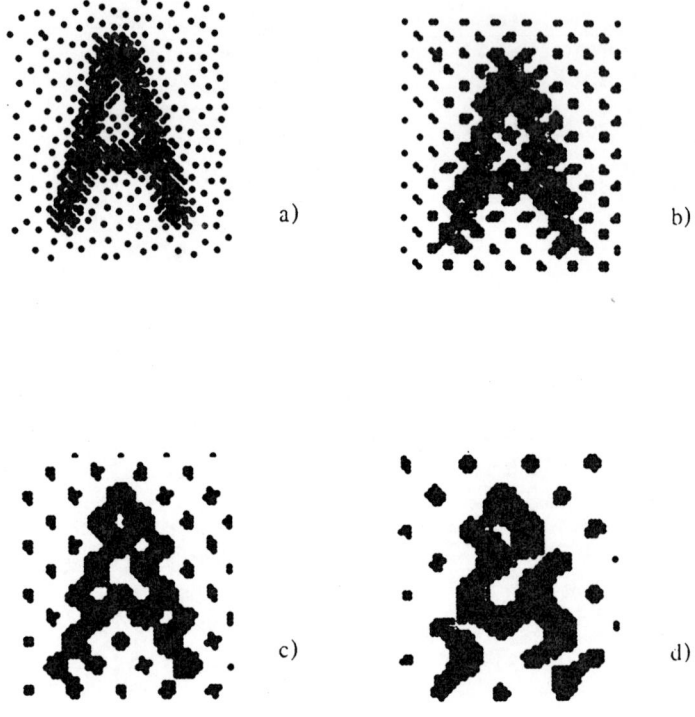

Figure 22. Cropped and magnified letter 'A'.

    a) MECCA, Q=256.
    b) SUPER-CIRCLE, Q=32.
       Screening resolution: 70 clusters of dots/inch.
    c) SUPER-CIRCLE, Q=64.
       Screening resolution: 47 clusters of dots/inch.
    d) SUPER-CIRCLE, Q=128.
       Screening resolution: 35 clusters of dots/inch.

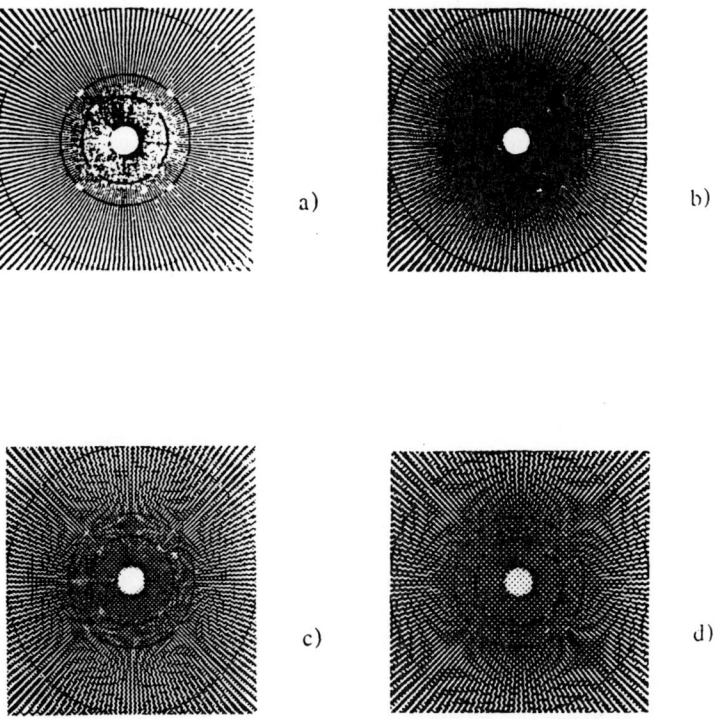

Figure 23. Bi-level representation of continuous-tone graphics.

a) MECCA, Q=256.
b) SUPER-CIRCLE, Q=32.
   Screening resolution: 70 clusters of dots/inch.
c) SUPER-CIRCLE, Q=64.
   Screening resolution: 47 clusters of dots/inch.
d) SUPER-CIRCLE, Q=128.
   Screening resolution: 35 clusters of dots/inch.

graphic structure with an ordered dot-pattern texture as encountered in SUPER-CIRCLE halftoning, makes the latter a poor choice for computer-output printing of continuous-tone graphics. This experiment also demonstrated the higher-resolution rendition capability and the Moiré-pattern formation inhibition property of MECCA, features of primary importance in all two-dimensional resampling tasks, i.e., scanning, interpolation, decimation, etc.

Continuous-tone natural image (Figure 24)

Due to the dot-size correction capability of MECCA, continuous-tone natural images, where tonal information is known to be visually important, is reproduced with a minimum loss or shift of tonal gradation and without noticeable screening artifacts. In contrast, the visibility of a slight overdarkening due to circular-shaped print dots, leading to excessive area coverage along the envelope of clusters of dots in SUPER-CIRCLE screening, again makes the latter an inferior choice for computer-output printing of continuous-tone natural images. Note that the overdarkening effect is reduced, when using lower-resolution screens. At the same time, however, subjectively disturbing masking due to larger clusters of dots gets increased.

Independent of picture content, the subjective appearance of MECCA-processed bi-level halftone representations is far superior in quality to the one obtained with SUPER-CIRCLE screening. This improved economy of representation is achieved at the expense of increased computational complexity. While the heavy I/O traffic encountered in non-coded information-processing applications, is handled with similar overhead in MECCA and in SUPER-CIRCLE screening, the actual signal-processing tasks to be performed differ substantially from one technique to the other.

The program kernels in MECCA primarily perform *convolutions and table look-ups,* the ones in SUPER-CIRCLE screening *comparisons and cyclic rearrangements* of threshold values. In both cases, the computation time is a linear function of the number of pels processed. The execution-time ratio between MECCA and SUPER-CIRCLE screening software, implemented as PL/I programs with embedded assembler subroutines and run on various IBM/370 main-frame computers, is approximately 5:1. On a $0.8 \cdot 10^6$ instruction-per-second general-purpose computer, the average execution time for MECCA-processed pels is approximately 80 μs; a figure which can be substantially reduced using VLSI-based special-purpose low-cost processor hardware.

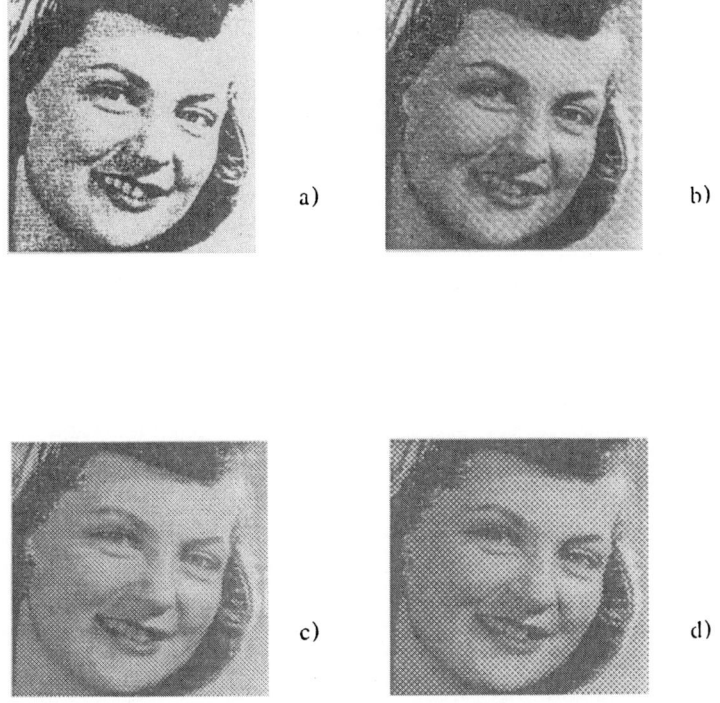

a)

b)

c)

d)

Figure 24. Bi-level representation of continuous-tone natural images.

    a) MECCA, Q=256.
    b) SUPER-CIRCLE, Q=32.
       Screening resolution: 70 clusters of dots/inch.
    c) SUPER-CIRCLE. Q=64.
       Screening resolution: 47 clusters of dots/inch.
    d) SUPER-CIRCLE, Q=128.
       Screening resolution: 35 clusters of dots/inch.

To obtain a better understanding of the various reproduction-quality versus computational-complexity trade-offs involved, a sufficiently *quantitative* rather than a solely *qualitative* investigation is necessary to assess the performance of MECCA and SUPER-CIRCLE screening techniques.

Contrast-Sensitivity Experiment for Performance Evaluation

In SUPER-CIRCLE screening, the quality of bi-level reproduction is limited by the fineness of the clusters of dots used. In MECCA, the quality depends on the fineness of the spatial extent over which a perceptually acceptable accuracy of the minimum average value can be reached. In general, the quality of an image reproduction process depends on its capability to reproduce spatial frequencies, and a possible solution to quantitatively evaluate its performance is to determine its modulation transfer function. The latter can be measured by generating frequency and amplitude-controlled sinusoidal gratings as shown in Figure 25.

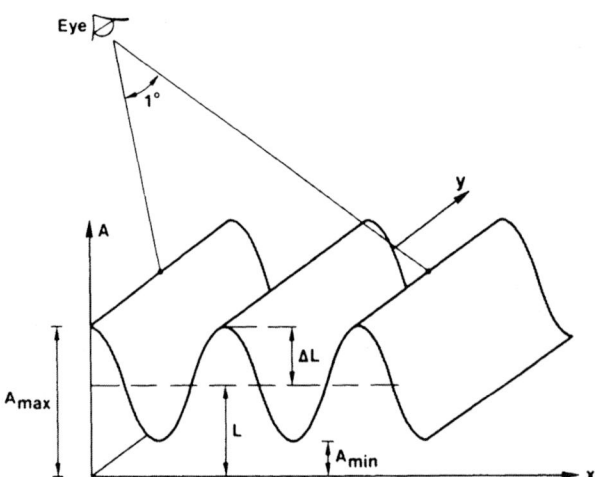

Figure 25. Parameters of sinusoidal grating.

The spatial frequency u of a grating is defined as the number of cycles that subtend an angle of one degree at the eye of its observer. The contrast or depth of modulation is defined as

$$C = (A_{max} - A_{min})/(A_{max} + A_{min}).$$

Figure 26 shows a MECCA and a SUPER-CIRCLE-processed bi-level representation of a grating in which the spatial frequency u and the reciprocal of the contrast $1/C$ vary logarithmically. As $1/C$ increases, the grating becomes more and more difficult to see until finally it appears as a uniform gray field. The intermediate zone, where the grating reaches threshold, is measurable by psychometric techniques, and its value is termed the *threshold for the perception of contrast or threshold contrast* C_T. Finally, the *contrast sensitivity* is defined as the reciprocal of C_T [10].

For more accurate measurement of the threshold contrast C_T, a large set of single spatial-frequency, single contrast-value gratings was generated using MECCA and SUPER-CIRCLE screening techniques and reproduced on a 400 pel/inch high-resolution computer-output printing device. The 2.5 x 2.5 inch^2 sized hardcopy stimuli obtained were then presented, in random order, to individual observers under a viewing box that provided a high-level, constant illumination and constrained the binocular viewing distance to 12.5 inch. To determine the contrast threshold, the observers were then asked to indicate for each vertically invariant and horizontally sinusoidally varying MECCA and SUPER-CIRCLE stimulus whether, yes or no, they could see any grating. Finally, the answers obtained from 15 observers were averaged and the reciprocal plotted as the contrast sensitivity of halftoned sine-wave gratings as a function of spatial frequency (Figure 27).

The contrast-sensitivity experiment described above is based on the properties of the human eye. In order to minimize the influence of unknown non-linear effects which might arise in a psychophysical experiment involving human vision, the performance evaluation of MECCA and SUPER-CIRCLE screening is based on a comparative interpretation. The results plotted in Figure 27 show that the eye perceives peak contrast sensitivities at a spatial frequency of about 1 cycle/degree for both MECCA and SUPER-CIRCLE halftoned sine-wave gratings. Furthermore, the contrast sensitivities perceived fall off at both higher and lower spatial frequencies. However, relative to each other, the peak contrast sensitivity observed for MECCA is twice as high as for SUPER-CIRCLE screening. Similarly, cut-off is reached for MECCA at

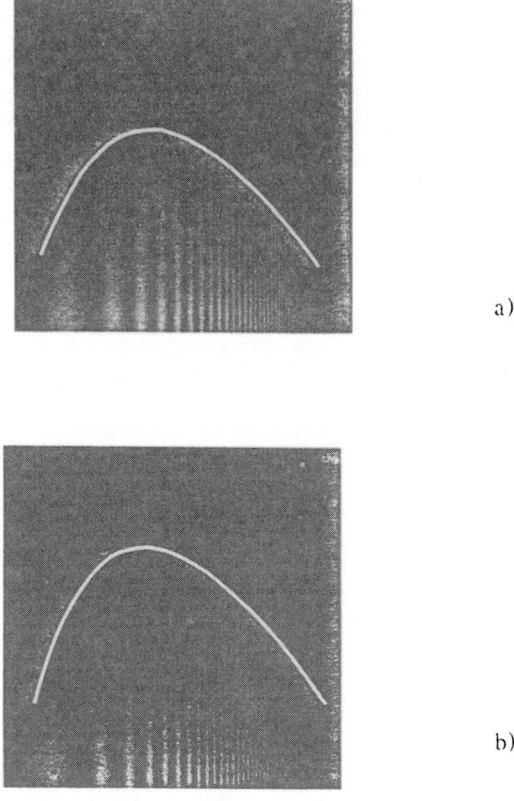

a)

b)

Figure 26. Bi-level representation of sinusoidal gratings showing the reciprocal of contrast as a function of spatial frequency.

a) SUPER-CIRCLE screening
(average gray-level set at $Q/2=32$).
b) MECCA
(average gray-value set at $Q/2=128$).

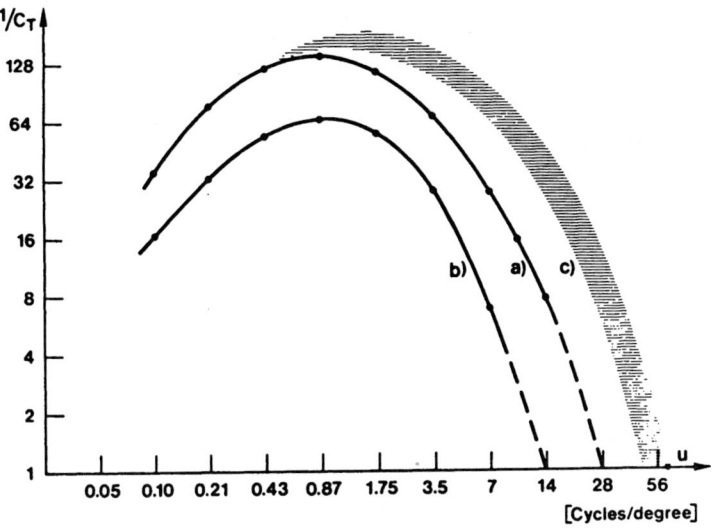

Figure 27. Observed contrast sensitivity of halftoned sine-wave
gratings as a function of spatial frequency.

a) MECCA.
b) SUPER-CIRCLE screening.
c) Contrast-sensitivity domain of the human eye [10].

28 cycles/degree, a spatial frequency twice as high as for SUPER-CIRCLE
screening. In order to adjust the performance of the latter to the level of
MECCA, the number of reproduced dots per unit area has to be quadrupled.
Consequently, the linear resolution of the computer-output printing device
has to be doubled and the execution-time ratio between MECCA and
SUPER-CIRCLE screening will fall to 5:4. Clearly, this is a bad engineering
trade-off, since the cost of computing is constantly decreasing while the ex-
penses for mechanical accuracy are steadily increasing.

The results plotted in Figure 27 further indicate that any spatial-frequency components that fall below threshold contrast C_T will not be observable by the average viewer and therefore need not be scanned and stored for bi-level halftone representation. For very high spatial frequencies, the threshold contrast C_T is high enough for all signals to be below this level. In practice, high spatial frequencies can be eliminated by a low enough resolution scanner or simply by low-pass filtering the scanned signal. A proper match between scanning and printing resolution also optimizes the data compression performance of continuous-tone compression schemes.

A perceptually 'perfect' bi-level halftone representation is one that ideally matches the performance of the human visual system. Figure 27 also exhibits the contrast-sensitivity domain of the human eye. Its exact limitation is difficult to indicate as it depends on many parameters such as pupil diameter, illumination and paper-reflectance conditions, etc. Without considering any technical and economic constraints, the contrast-sensitivity experiment shows that for normal viewing conditions, perceptually 'perfect' bi-level halftone representation can be achieved with MECCA and a 600 to 800 pel/inch resolution computer-output printing device.

Extensions of MECCA

The overall appearance of a bi-level image hardcopy reproduction is heavily affected by the quality of the individually printed dots. MECCA described here assumes a toner/ink deposition process capable of producing circular dots, characterized by uniform black coverage and sharp contours. Such hard dots can be achieved with various printing technologies and high-quality paper characterized by good surface smoothness and uniform ink receptivity.

For lower-quality paper or certain lower-cost printing technologies, the toner/ink deposition process tends to produce soft dots characterized by non-uniform black coverage and weak contours. In this case, the computation of the dot-size correction factor $\lambda(x,y)$ needs to be based on new design criteria.

The concept of multiple-error correction can also be extended, at the expense of increased computational complexity, to include further terms to compensate paper-specific shortcomings. For example, to neutralize the scattering of light inside dot-covered paper before it emerges, causing a slightly reduced reflectance over that expected when the internal scattering is not taken into account [11].

Application of MECCA

The concept of MECCA can be adapted to any system that incorporates all-point-addressable printers and displays of different resolution and technology. Although a detailed architectural presentation is beyond the scope of this paper, the conceptual appearance of such a system to the user can be described as follows: All stored or transmitted continuous- and halftone images are quantized for optimum image-quality rendition with an ideal reproduction device, i.e., assuming a linear relationship between the area coverage c and the amplitude quantization level q. The implementation of such a device independency concept is to add a header to each image file containing ORIGIN, DESTINATION and OUTPUT LOGICAL DEVICE attributes. If a reproduction OUTPUT LOGICAL DEVICE is specified, the image file will be put into temporary storage and automatically MECCA-processed with the set of parameters proper to the reproduction device specified. If no reproduction OUTPUT LOGICAL DEVICE is specified, the image file will be put into temporary storage and the receiver notified of its existence through higher-level functions such as data-base managers and message displays. The user can then specify the reproduction device and submit MECCA with the proper parameters.

ACKNOWLEDGMENT

The author would like to thank B. P. Medoff for early discussions on MECCA, H. Thomas and W. Butera for their treatment of dot-overlap geometries, P. Andreae for carrying out the psychophysical experiments, and C. d'Heureuse for preparing some of the halftone pictures.

APPENDIX I

Determination of $A_{overlap}(x,y)$, $A_{eff}(x,y)$ *and* $\lambda(x,y)$ *for* $L=4$

The total area of overlap at P(x,y) resulting from any combination of previously printed dots at P(x-1,y), P(x,y+1), P(x-1,y+1) and P(x+1,y+1) can be reconstructed from five component blocks F1, F2, F3, F4 and F5 as shown in Figure Ia).

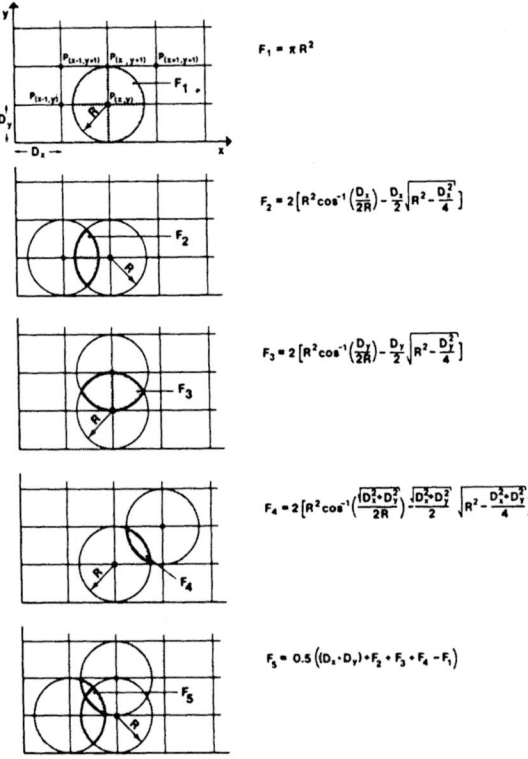

$$F_1 = \pi R^2$$

$$F_2 = 2\left[R^2\cos^{-1}\left(\frac{D_x}{2R}\right) - \frac{D_x}{2}\sqrt{R^2 - \frac{D_x^2}{4}}\right]$$

$$F_3 = 2\left[R^2\cos^{-1}\left(\frac{D_y}{2R}\right) - \frac{D_y}{2}\sqrt{R^2 - \frac{D_y^2}{4}}\right]$$

$$F_4 = 2\left[R^2\cos^{-1}\left(\frac{\sqrt{D_x^2+D_y^2}}{2R}\right) - \frac{\sqrt{D_x^2+D_y^2}}{2}\sqrt{R^2 - \frac{D_x^2+D_y^2}{4}}\right]$$

$$F_5 = 0.5\left((D_x \cdot D_y) + F_2 + F_3 + F_4 - F_1\right)$$

Figure Ia). The five component blocks to compute the area of overlap.

An example demonstrating the computation of $A_{overlap}(x,y)$ is shown in Figure 1b).

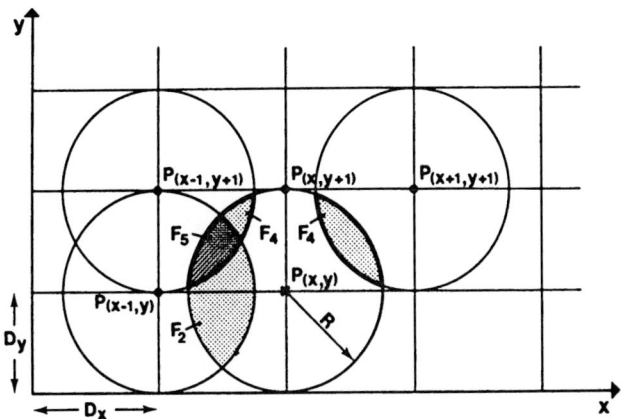

Figure 1b). Example of overlap computation using component blocks.

For given parameters D_x, D_y and R, the overlap between the circles centered at $P(x,y)$ and $P(x+1,y+1)$ is given by F4. The overlap between the circles centered at $P(x,y)$ and $P(x-1,y+1)$ is also described by F4. The area common to the circles centered at $P(x,y)$ and $P(x-1,y)$ is given by F2. Finally, the common area between three mutually adjacent dots centered at $P(x,y)$, $P(x-1,y)$ and $P(x-1,y+1)$ is described by F5. The total area of overlap therefore amounts to

$$A_{overlap}(x,y) = F2 + 2F4 - F5.$$

The effective area of white paper newly covered is

$$A_{eff}(x,y) = F1 - A_{overlap}(x,y),$$

and the dot-size correction value

$$\lambda(x,y) = A_{eff}(x,y)/(D_x \, D_y).$$

Table 1 lists all possible overlap combinations for $L=4$ adjacent print positions and the corresponding formulas to compute $A_{eff}(x,y)$.

P(x+1,y+1)	P(x,y+1)	P(x-1,y+1)	P(x-1,y)	$A_{eff}(x,y)$
0	0	0	0	I1
0	0	0	1	I1 - I2
0	0	1	0	I1 - I4
0	0	1	1	I1 - I2 - I4 + I5
0	1	0	0	I1 - I3
0	1	0	1	I1 - I2 - I3 + I5
0	1	1	0	I1 - I3 - I4 + I5
0	1	1	1	I1 - I2 - I3 + I5
1	0	0	0	I1 - I4
1	0	0	1	I1 - I4 - I2
1	0	1	0	I1 - 2I4
1	0	1	1	I1 - I2 - 2I4 + I5
1	1	0	0	I1 - I3 - I4 + I5
1	1	0	1	I1 - I2 - I3 - I4 + 2I5
1	1	1	0	I1 - I3 - 2I4 + 2I5
1	1	1	1	I1 - I2 - I3 - I4 + 2I5

('1' = dot, '0' = no dot)

Table I. All possible overlap contributions for L=4 adjacent print positions and the corresponding expressions for $A_{eff}(x,y)$.

APPENDIX II

Determination of $A_{overlap}(x,y)$, $A_{eff}(x,y)$ *and* $\lambda(x,y)$ *for* $L=6$

The total area of overlap at $P(x,y)$ resulting from any combination of previously printed dots at $P(x-1,y)$, $P(x,y+1)$, $P(x-1,y+1)$, $P(x+1,y+1)$, $P(x-2,y)$ and $P(x,y+2)$ can be reconstructed using the technique of component blocks. For $L=6$ dots, nine component blocks are necessary to reconstruct the total overlap for a given printed dot configuration. Five of these blocks are the same as the ones used in the $L=4$ dot case, while the remaining four blocks are unique to the $L=6$ dot case [Figure IIa)].

An example involving $L=6$ potentially overlapping neighbor dots is shown in Figure IIb). Here, the dots centered at print positions $P(x-1,y+1)$, $P(x,y+1)$ and $P(x-2,y)$ are already present when the dot centered at $P(x,y)$ is printed.

In order to compute the total overlap area $A_{overlap}(x,y)$, the contribution from each printed neighboring dot is considered and adjustments are made to ensure that no part of an overlap is counted more than once. For the case shown in Figure IIb),

$$A_{eff}(x,y) = F1 - A_{overlap}(x,y) = F1 - [F3 + (F4 - F5) + (F6 - F8)],$$

where

F1 :	area of circular dot,	
F3 :	overlap between dot centered at $P(x,y)$ and dot centered at $P(x,y+1)$,	
F4 :	overlap between dot centered at $P(x,y)$ and dot centered at $P(x-1,y+1)$,	
F5 :	correction value used to keep the area common to dots centered at $P(x,y+1)$, $P(x-1,y+1)$ and $P(x,y)$ from being counted twice,	
F6 :	overlap between dot centered at $P(x,y)$ and dot centered at $P(x-2,y)$,	
F8 :	correction value used to keep the area common to dots centered at $P(x-1,y+1)$, $P(x-2,y)$ and $P(x,y)$ from being counted twice.	

Table II lists all possible overlap combinations for $L=6$ adjacent print positions and the corresponding formulas to compute $A_{eff}(x,y)$.

P(x,x+2)		P(x+1,x+1)		P(x-1,x+3)		A_eff(x,y)
	P(x-2,x)		P(x,x+1)		P(x,x)	
0	0	0	0	0	0	I1
0	0	0	0	0	1	I1 + I2
0	0	0	0	1	0	I1 + I4
0	0	0	0	1	1	I1 + I2 + I4 + I5
0	0	0	1	0	0	I1 + I3
0	0	0	1	0	1	I1 + I2 + I3 + I5
0	0	0	1	1	0	I1 + I3 + I4 + I5
0	0	0	1	1	1	I1 + I2 + I3 + I5
0	0	1	0	0	0	I1 + I4
0	0	1	0	0	1	I1 + I4 + I5
0	0	1	0	1	0	I1 + 2I4 + I8
0	0	1	0	1	1	I1 + I2 + I4 + I5 + I8
0	0	1	1	0	0	I1 + I4 + I3 + I5
0	0	1	1	0	1	I1 + I4 + I3 + I2 + 2I5
0	0	1	1	1	0	I1 + 2I4 + I3 + 2I5
0	0	1	1	1	1	I1 + I4 + I3 + I2 + 2I5
0	1	0	0	0	0	I1 + I6
0	1	0	0	0	1	I1 + I2
0	1	0	0	1	0	I1 + I6 + I4 + I8
0	1	0	0	1	1	I1 + I2 + I4 + I5
0	1	0	1	0	0	I1 + I6 + I5
0	1	0	1	0	1	I1 + I2 + I3 + I5
0	1	0	1	1	0	I1 + I3 + I4 + I5 + I8 + I6
0	1	0	1	1	1	I1 + I3 + I4 + I5
0	1	1	0	0	0	I1 + I6 + I4
0	1	1	0	0	1	I1 + I4 + I5
0	1	1	0	1	0	I1 + I6 + I4 + I8 + I4 + I8
0	1	1	0	1	1	I1 + I2 + 2I4 + I5 + I8
0	1	1	1	0	0	I1 + I6 + I3 + I4 + I5
0	1	1	1	0	1	I1 + I4 + I3 + I2 + 2I5
0	1	1	1	1	0	I1 + I6 + I4 + I8 + I3 + I4 + 2I5
0	1	1	1	1	1	I1 + I4 + I3 + I2 + 2I5
1	0	0	0	0	0	I1 + I3
1	0	0	0	0	1	I1 + I2 + I3
1	0	0	0	1	0	I1 + I4 + I3 + I9
1	0	0	0	1	1	I1 + I2 + I4 + I5 + I3 + I9
1	0	0	1	0	0	I1 + I3
1	0	0	1	0	1	I1 + I2 + I3 + I5
1	0	0	1	1	0	I1 + I3 + I2 + I5
1	0	0	1	1	1	I1 + I2 + I3 + I5
1	0	1	0	0	0	I1 + I4 + I3 + I9
1	0	1	0	0	1	I1 + I3 + I4 + I3 + I9
1	0	1	0	1	0	I1 + 2I4 + I8
1	0	1	0	1	1	I1 + I2 + 2I4 + I5 + I8
1	0	1	1	0	0	I1 + I4 + I3 + I5
1	0	1	1	0	1	I1 + I4 + I3 + I2 + 2I5
1	0	1	1	1	0	I1 + 2I4 + I3 + 2I5
1	0	1	1	1	1	I1 + I4 + I3 + I2 + 2I5
1	1	0	0	0	0	I1 + I6 + I3
1	1	0	0	0	1	I1 + I2 + I3
1	1	0	0	1	0	I1 + I3 + I6 + I4 + I8 + I9
1	1	0	0	1	1	I1 + I2 + I4 + I5 + I3 + I9
1	1	0	1	0	0	I1 + I6 + I3
1	1	0	1	0	1	I1 + I2 + I3 + I5
1	1	0	1	1	0	I1 + I3 + I4 + I5 + I8 + I6
1	1	0	1	1	1	I1 + I3 + I3 + I5
1	1	1	0	0	0	I1 + I4 + I3 + I9 + I6
1	1	1	0	0	1	I1 + I2 + I4 + I3 + I9
1	1	1	0	1	0	I1 + I6 + I4 + I8 + I3 + I4 + I8
1	1	1	0	1	1	I1 + I2 + 2I4 + I5 + I8
1	1	1	1	0	0	I1 + I6 + I3 + I4 + I5
1	1	1	1	0	1	I1 + I4 + I3 + I2 + 2I5
1	1	1	1	1	0	I1 + I6 + 2I4 + I8 + I3 + 2I5
1	1	1	1	1	1	I1 + I4 + I3 + I2 + 2I5

('1' = dot, '0' = no dot)

Table II. All possible overlap contributions for I = 6 adjacent print positions and the corresponding expressions for A_eff(x,y).

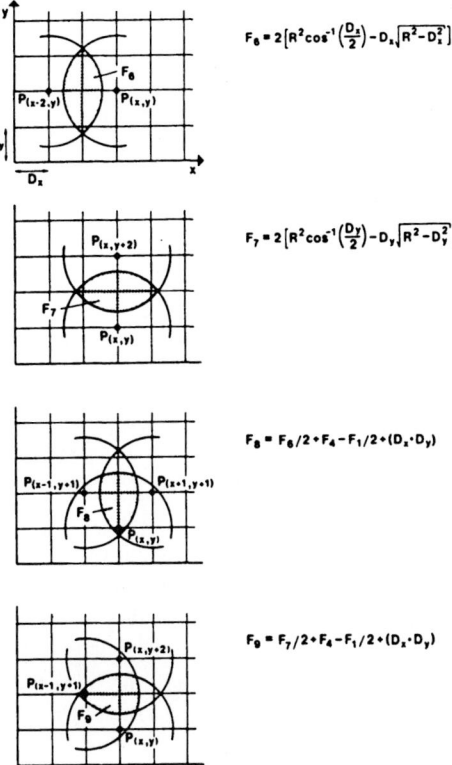

$$F_6 = 2\left[R^2\cos^{-1}\left(\frac{D_x}{2}\right) - D_x\sqrt{R^2 - D_x^2}\right]$$

$$F_7 = 2\left[R^2\cos^{-1}\left(\frac{D_y}{2}\right) - D_y\sqrt{R^2 - D_y^2}\right]$$

$$F_8 = F_6/2 + F_4 - F_1/2 + (D_x \cdot D_y)$$

$$F_9 = F_7/2 + F_4 - F_1/2 + (D_x \cdot D_y)$$

Figure 11a). The four component blocks unique to the $L=6$ dot case.

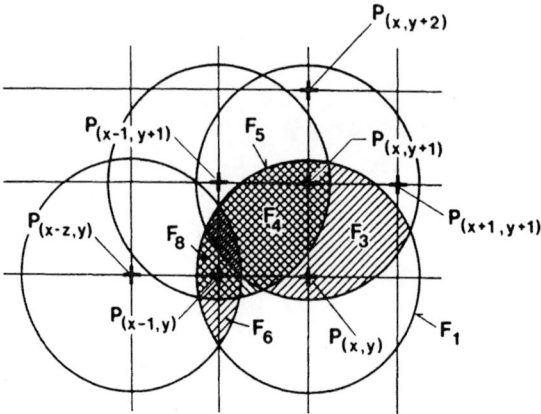

Figure 11b). Example involving L=6 potentially overlapping
neighbor dots.

REFERENCES

[1] P. Stucki, 'Statistical Measurements for Television Picture Classi-
fication and Some Experiments in Video Signal Encoding', DIC
Thesis, Imperial College of Science and Technology, London, July
1968.

[2] J. O. Limb, 'Design of Dither Waveforms for Quantized Visual
Signals', Bell Syst. Tech. J., Vol. 48, No. 7, 1969.

[3] B. E. Bayer, 'An Optimum Method for Two-Level Rendition of
Continuous-Tone Pictures', International Conference on Communications, Conference Record, CAT. No. 73 CHO 744-3CSCB, June
1973.

[4] P. Stucki, 'Comparison and Optimization of Computer-Generated
Digital Halftone Pictures', SID Digest of Technical Papers, Vol. VI,
1975.

[5] G. C. Higgins and K. Stultz, 'Visual Acuity as Measured with
Various Orientations of a Parallel-Line Test Object', J. Opt. Soc.
Am., Vol. 38, No. 9, 1948.

[6] P. Stucki, 'Image Processing for Document Reproduction' in
Advances in Digital Image Processing, P. Stucki, Ed., Plenum
Presss, NY, 1979.

[7] K. Y. Wong and P. Stucki, 'Adaptive Switching of Dispersed and
Clustered Dot Patterns for Bi-Level Halftone Rendition', SID Digest
of Technical Papers, Vol. VIII, 1977.

[8] R. Floyd, 'An Adaptive Algorithm for Spatial Gray-Scale', SID Digest
of Technical Papers, Vol. VI, 1975.

[9] J. F. Jarvis, C. N. Judice and W. H. Ninke, 'A Survey of Techniques
for the Display of Continuous-Tone Pictures on Bilevel Displays',
Computer Graphics and Image Processing, Vol. 5, No. 1, 1976.

[10] F. W. Campbell, 'The Human Eye as an Optical Filter', Proc. IEEE,
Vol. 56, No. 6, 1968.

[11] F. R. Clapper and J. A. C. Yule, 'The Effect of Multiple Internal
Reflections on the Densities of Half-Tone Prints on Paper', J. Opt.
Soc. Am., Vol. 43, No. 7, 1953.

Author Index

The page numbers refer to the list of references provided by each contributor.

Vol. 117: Fundamentals of Computation Theory. Proceedings, 1981. Edited by F. Gécseg. XI, 471 pages. 1981.

Vol. 118: Mathematical Foundations of Computer Science 1981. Proceedings, 1981. Edited by J. Gruska and M. Chytil. XI, 589 pages. 1981.

Vol. 119: G. Hirst, Anaphora in Natural Language Understanding: A Survey. XIII, 128 pages. 1981.

Vol. 120: L. B. Rall, Automatic Differentiation: Techniques and Applications. VIII, 165 pages. 1981.

Vol. 121: Z. Zlatev, J. Wasniewski, and K. Schaumburg, Y12M Solution of Large and Sparse Systems of Linear Algebraic Equations. IX, 128 pages. 1981.

Vol. 122: Algorithms in Modern Mathematics and Computer Science. Proceedings, 1979. Edited by A. P. Ershov and D. E. Knuth. XI, 487 pages. 1981.

Vol. 123: Trends in Information Processing Systems. Proceedings, 1981. Edited by A. J. W. Duijvestijn and P. C. Lockemann. XI, 349 pages. 1981.

Vol. 124: W. Polak, Compiler Specification and Verification. XIII, 269 pages. 1981.

Vol. 125: Logic of Programs. Proceedings, 1979. Edited by E. Engeler. V, 245 pages. 1981.

Vol. 126: Microcomputer System Design. Proceedings, 1981. Edited by M. J. Flynn, N. R. Harris, and D. P. McCarthy. VII, 397 pages. 1982.

Voll. 127: Y.Wallach, Alternating Sequential/Parallel Processing. X, 329 pages. 1982.

Vol. 128: P. Branquart, G. Louis, P. Wodon, An Analytical Description of CHILL, the CCITT High Level Language. VI, 277 pages. 1982.

Vol. 129: B. T. Hailpern, Verifying Concurrent Processes Using Temporal Logic. VIII, 208 pages. 1982.

Vol. 130: R. Goldblatt, Axiomatising the Logic of Computer Programming. XI, 304 pages. 1982.

Vol. 131: Logics of Programs. Proceedings, 1981. Edited by D. Kozen. VI, 429 pages. 1982.

Vol. 132: Data Base Design Techniques I: Requirements and Logical Structures. Proceedings, 1978. Edited by S.B. Yao, S.B. Navathe, J.L. Weldon, and T.L. Kunii. V, 227 pages. 1982.

Vol. 133: Data Base Design Techniques II: Proceedings, 1979. Edited by S.B. Yao and T.L. Kunii. V, 229–399 pages. 1982.

Vol. 134: Program Specification. Proceedings, 1981. Edited by J. ᴀunstrup. IV, 426 pages. 1982.

Vol. 135: R.L. Constable, S.D. Johnson, and C.D. Eichenlaub, An Introduction to the PL/CV2 Programming Logic. X, 292 pages. 1982.

Vol. 136: Ch. M. Hoffmann, Group-Theoretic Algorithms and Graph Isomorphism. VIII, 311 pages. 1982.

Vol. 137: International Symposium on Programming. Proceedings, 1982. Edited by M. Dezani-Ciancaglini and M. Montanari. VI, 406 pages. 1982.

Vol. 138: 6th Conference on Automated Deduction. Proceedings, 1982. Edited by D.W. Loveland. VII, 389 pages. 1982.

Vol. 139: J. Uhl, S. Drossopoulou, G. Persch, G. Goos, M. Dausmann, G. Winterstein, W. Kirchgässner, An Attribute Grammar for the Semantic Analysis of Ada. IX, 511 pages. 1982.

Vol. 140: Automata, Languages and programming. Edited by M. Nielsen and E.M. Schmidt. VII, 614 pages. 1982.

Vol. 141: U. Kastens, B. Hutt, E. Zimmermann, GAG: A Practical Compiler Generator. IV, 156 pages. 1982.

Vol. 142: Problems and Methodologies in Mathematical Software Production. Proceedings, 1980. Edited by P.C. Messina and A. Murli. VII, 271 pages. 1982.

Vol. 143: Operating Systems Engineering. Proceedings, 1980. Edited by M. Maekawa and L.A. Belady. VII, 465 pages. 1982.

Vol. 144: Computer Algebra. Proceedings, 1982. Edited by J. Calmet. XIV, 301 pages. 1982.

Vol. 145: Theoretical Computer Science. Proceedings, 1983. Edited by A.B. Cremers and H.P. Kriegel. X, 367 pages. 1982.

Vol. 146: Research and Development in Information Retrieval. Proceedings, 1982. Edited by G. Salton and H.-J. Schneider. IX, 311 pages. 1983.

Vol. 147: RIMS Symposia on Software Science and Engineering. Proceedings, 1982. Edited by E. Goto, I. Nakata, K. Furukawa, R. Nakajima, and A. Yonezawa. V. 232 pages. 1983.

Vol. 148: Logics of Programs and Their Applications. Proceedings, 1980. Edited by A. Salwicki. VI, 324 pages. 1983.

Vol. 149: Cryptography. Proceedings, 1982. Edited by T. Beth. VIII, 402 pages. 1983.

Vol. 150: Enduser Systems and Their Human Factors. Proceedings, 1983. Edited by A. Blaser and M. Zoeppritz. III, 138 pages. 1983.

Vol. 151: R. Piloty, M. Barbacci, D. Borrione, D. Dietmeyer, F. Hill, and P. Skelly, CONLAN Report. XII, 174 pages. 1983.

Vol. 152: Specification and Design of Software Systems. Proceedings, 1982. Edited by E. Knuth and E.J. Neuhold. V, 152 pages. 1983.

Vol. 153: Graph-Grammars and Their Application to Computer Science. Proceedings, 1982. Edited by H. Ehrig, M. Nagl, and G. Rozenberg. VII, 452 pages. 1983.

Vol. 154: Automata, Languages and Programming. Proceedings, 1983. Edited by J. Díaz. VIII, 734 pages. 1983.

Vol. 155: The Programming Language Ada. Reference Manual. Approved 17 February 1983. American National Standards Institute, Inc. ANSI/MIL-STD-1815A-1983. IX, 331 pages. 1983.

Vol. 156: M. H. Overmars, The Design of Dynamic Data Structures. VII, 181 pages. 1983.

Vol. 157: O. Østerby, Z. Zlatev, Direct Methods for Sparse Matrices. VIII, 127 pages. 1983.

Vol. 158: Foundations of Computation Theory. Proceedings, 1983. Edited by M. Karpinski, XI, 517 pages. 1983.

Vol. 159: CAAP'83. Proceedings, 1983. Edited by G. Ausiello and M. Protasi. VI, 416 pages. 1983.

Vol. 160: The IOTA Programming System. Edited by R. Nakajima and T. Yuasa. VII, 217 pages. 1983.

Vol. 161: DIANA, An Intermediate Language for Ada. Edited by G. Goos, W.A. Wulf, A. Evans, Jr. and K.J. Butler. VII, 201 pages. 1983.

Vol. 162: Computer Algebra. Proceedings, 1983. Edited by J.A. van Hulzen. XIII, 305 pages. 1983.

Vol. 163: VLSI Engineering. Proceedings. Edited by T.L. Kunii. VIII, 308 pages. 1984.